Zeitschrift für Geomorphologie

Annals of Geomorphology
Annales de Géomorphologie

A journal recognized by the International Association of Geomorphologists (IAG)

Neue Folge
Supplement Volume **132**

Geophysical Applications in Geomorphology

edited by Lothar Schrott, Andreas Hördt and Richard Dikau

with 72 figures and 15 tables in the text
and 1 CD

Gebrüder Borntraeger · Berlin · Stuttgart 2003

ISBN 3-443-21132-1 / ISSN 0044-2798
Sonderheft zur Zeitschrift für Geomorphologie, N. F., Band 47, 2003
Verlag: Gebrüder Borntraeger, Johannesstraße 3 A, D-70176 Stuttgart, http://www.schweizerbart.de
© by Gebrüder Borntraeger, Berlin · Stuttgart, 2003
∞ Printed on permanent paper conforming to ISO 9706 – 1994
All rights reserved including translation into foreign languages. This journal or parts thereof may not be reproduced in any form without permission from the publishers.
Valid for users in USA: The appearance of the code at the bottom of the first page of an article in this journal indicates the copyright owner's consent that copies of the article may be made for personal or internal use, or for the personal or internal use of specific clients. This consent is given on the condition, however, that the copier pays the stated per-copy fee through the Copyright Clearance Center, Inc., P.O.B. 8891, Boston, Mass. 02114, for copying beyond that permitted by Sections 107 or 108 of the Copyright Law.
Printed in Germany by K. Triltsch, Print und digitale Medien GmbH, Ochsenfurt-Hohestadt

Contents

Schrott, L. & Hördt, A.: Preface: Geophysical Applications in Geomorphology . V–VII

Theoretical issue

Ballantyne, C. K.: Paraglacial landform succession and sediment storage in deglaciated mountain valleys: theory and approaches to calibration (with 6 figures) ... 1– 18

DC-resistivity

Etzelmüller, B., Berthling, I. & Ødegard, R. S.: One-dimensional DC-resistivity depth soundings as a tool in permafrost investigations in high mountain areas of Southern Norway (with 8 figures and 1 table) 19– 36

Kneisel, Ch.: Electrical resistivity tomography as a tool for geomorphological investigations – some case studies (with 7 figures and 1 table) 37– 49

Sass, O.: Moisture distribution in rockwalls derived from 2D-resistivity measurements (with 11 figures and 3 tables) 51– 69

Seismic refraction

Hoffmann, Th. & Schrott, L.: Determining sediment thickness of talus slopes and valley fill deposits using seismic refraction – a comparison of 2D interpretation tools (with 7 figures and 3 tables) 71– 87

Hecht, St.: Differentiation of loose sediments with seismic refraction methods – potentials and limitations derived from case studies (with 11 figures) 89–102

Other methods

Berthling, I., Etzelmüller, B., Wåle, M. & Sollid, J. L.: Use of Ground Penetration Radar (GPR) soundings for investigating internal structures in rock glaciers. Examples from Prins Karls Forland, Svalbard (with 5 figures and 4 tables) 103–121

Hördt, A. & Zacher, G.: The radiomagnetotelluric method and its potential application in geomorphology (with 10 figures) 123–143

Combined applications

Kneisel, Ch. & Hauck, Ch.: Multi-method geophysical investigation of a sporadic permafrost occurrence (with 5 figures) 145–159

Hauck, Ch. & Vonder Mühll, D.: Evaluation of geophysical techniques for application in mountain permafrost studies (with 2 figures and 3 tables) 161–190

Addendum

CD-ROM: Demo-version Reflex version 3.0 (incl. handbook) and introduction to modelling and interpretation tools by K. J. Sandmeier

Preface: Geophysical applications in geomorphology

The use of geophysical techniques has become an important tool in many geomorphological studies. However, the correct handling of geophysical instruments and the subsequent processing of the data they yield, on the one hand, and the description and interpretation of geomorphological settings to which they are applied, on the other hand, are difficult tasks. Without a close cooperation of geophysicists and geomorphologists, the accurate and effective use of geophysical techniques and their geophysical and geomorphological interpretation is often limited. There are many text books in both disciplines, but no single book addresses the interdisciplinary aspects of combining geophysics and geomorphology. This special volume of "Zeitschrift für Geomorphologie" is intended to bridge the gap between geophysics and geomorphology and to encourage a closer cooperation between researchers in both disciplines. "Geophysical Applications in Geomorphology" is a collection of 10 papers covering a wide range of different geophysical techniques and their integration into geomorphological studies. Most of these papers were originally presented at the International Workshop "Geophysical Techniques in Geomorphology" held in Bonn, Germany in February 2002 with participants (geomorphologists and geophysicists) from Germany, Switzerland, United Kingdom and Norway. The objectives of the workshop were

- to present geomorphic studies demonstrating the powerful integration of geophysical techniques,
- to discuss and evaluate the efficiency of different geophysical techniques,
- to give a critical overview of modelling and software tools,
- to demonstrate the application of different geophysical instruments, and
- to exchange knowledge and to stimulate research groups with similar interests.

This volume addresses these objectives and illustrates the theoretical and practical issues of using geophysical techniques in geomorphology.

In the first paper Colin Ballantyne demonstrates the crucial integration of geophysical methods to determine more accurately sediment storage in order to calibrate models of paraglacial landform succession. He describes the theoretical bases for paraglacial sediment release in deglaciated mountain valleys, outlines the concept of paraglacial succession and develops a model of paraglacial sediment storage. The next section of three papers shows examples of DC-resistivity measurements in geomorphological research. The paper by Etzelmüller et al. gives a summary of 1D DC-resistivity soundings in southern Norway, where active layer thicknesses and permafrost occurrence is mapped based on different patterns of resistivity curves. The limitations of 1D resistivity soundings in permafrost mapping are clearly addressed. Christof Kneisel applied electrical resisitivity tomography to determine the structure of the subsurface at different study sites with high resistive, intermediate, and conductive material.

Special attention is also given to the choice of electrode configuration and geological and geomorphological interpretation of different subsurface resistivities. Oliver Sass describes a high resolution 2D-resistivity approach to determine moisture distribution in rockwalls, and provides insights to small scale moisture gradients, their temporal variation, and the implications for weathering.

In the next section, two examples of seismic refraction in geomorphological studies are presented. Thomas Hoffmann and Lothar Schrott apply four interpretation tools (intercept-time, wavefront-inversion, network-raytracing, and refraction tomography) to seismic data from two alpine locations. The authors discuss particular features of each interpretation tool and outline a possible interpretation strategy. A second paper on 2D seismic refraction by Stefan Hecht demonstrates the potential and limitation of this method with respect to the differentiation of loose sediments. In four different environments, the delimitation of loose sediments from bedrock is shown and problems such as those posed by hidden layers are addressed.

The next section deals with ground pentrating radar (rarely used in geomorphology previously) and radiomagnetotellurics (not yet applied to geomorphological problems).

Berthling et al. show the potential of ground pentrating radar (GPR) for investigating internal structures in rock glaciers. This study includes detailed descriptions of reflectors and their relation to shear zones and to the bedrock interface as well as recommendations for GPR studies on rock glaciers. Andreas Hördt and Gerd Zacher suggest the potential application of radiomagnetotelluric (RMT) in geomorphology. Based on three case studies the possible range for exploring groundwater, waste deposits, and small cavities is illustrated. Given highly portable instrumentation, RMT should be particularly useful for future geomorphological studies in inaccessible terrain such as alpine environments.

The final section of the issue includes two papers which combine geophysical applications in geomorphological research. Christof Kneisel and Christian Hauck present a multi-method geophysical approach involving one-dimensional DC resistivity soundings, two-dimensional DC resistivity tomography, seismic refraction, electromagnetics and capacitively coupled geoelectrical profiling. The methods were used to determine the existence of permafrost below treeline in Switzerland and were evaluated with respect to their applicability on mountain permafrost. In the last paper Christian Hauck and Daniel Vonder-Mühll undertake a comprehensive review of five different geophysical techniques and their application in mountain permafrost studies. They address five major types of application (general permafrost prospecting, mapping of lateral variation, determination of vertical extent, active layer studies and permafrost monitoring) and evaluate the advantages and disadvantages of different methods within each of these categories. They include DC resistivity tomography, refraction seismic tomography, electromagnetic induction methods, ground pentrating radar, and the bottom temperature of snow cover (BTS) in their review.

In addition to these papers, we include a CD-ROM by Karl-Josef Sandmeier which contains (i) a test version of the software package REFLEXW (programme for 2D and 3D processing and interpretation of GPR and seismic refraction/reflection data, incl. Demodata and handbook); (ii) an introduction to the interpretation of seismic refraction data; (iii) an introduction to modelling and tomography tools, and (iv) a technical note on the use of wavefront inversion, forward modelling and tomographic interpretation tools for seismic refraction data.

Finally we would like to thank Prof. Richard Dikau for stimulating discussions on current research problems on the intersection of both disciplines and for initiating the Workshop within the framework of the Graduiertenkolleg 437 "Landform – a structured and variable boundary surface". Financial support from the Deutsche Forschungsgemeinschaft (DFG) is gratefully acknowledged.

We thank all the colleagues who, in their role as referees, have guaranteed a high scientific level for all contributions collected in this issue. In particular we are grateful to: R. Bisdorf, N. Blindow, R.-U. Boerner, O. Boyd, N. Caine, M. Friberg, S. Friedel, K. Hall, B. Hobbs, L. King, J. Lucius, D. Orlowski, T. Pfeffer, B. Rodriguez, H. Rüter, A. Sheehan, B. Siemon, C. Thorn, and several anonymous referees.

We are also thankful to Prof. K.-H. Pfeffer (chief editor) and Dr. E. Nägele (Gebr. Borntraeger Verlagsbuchhandlung, Science Publishers) for giving us the possibility of publishing the results of the Workshop in this special volume of "Zeitschrift für Geomorphologie".

Lothar Schrott and Andreas Hördt, Bonn

Paraglacial landform succession and sediment storage in deglaciated mountain valleys: theory and approaches to calibration

Colin K. Ballantyne, St Andrews, Scotland

with 6 figures

Summary. Paraglacial landform succession reflects the fact that different paraglacial subsystems relax over different timescales, so that some paraglacial sediment stores experience net erosion even as others continue to accumulate. Over millennial timescales individual sediment stores (such as talus, debris cones, alluvial fans and valley fills) exhibit net sediment accumulation then net sediment loss. A general steady-state model of paraglacial sediment storage is developed based on the equation $S = (S_a - S_a e^{-\lambda t}) e^{-\kappa t}$, where t is time elapsed since deglaciation, S is storage volume at time t, S_a is the volume of 'available' sediment feeding the sediment store, λ is the rate of sediment release from the sediment source and κ is the rate of sediment loss from the sediment store. It is shown that sediment budgets for any time t can be retrodicted or predicted through derivation of S_a, λ and κ provided that two points on the curve described by the above equation are known: (t', S_m) and (t'', S_v), where t' is time since deglaciation when the sediment store reaches its maximum volume, S_m is the maximum volume of stored sediment, t'' is time elapsed since deglaciation and S_v is the present volume of stored sediment. Recent developments in the use of geophysical techniques to establish the present (and thereby former maximum) volumes of paraglacial sediment stores form an essential component of this approach, which has the potential to allow the changing sediment budget of individual paraglacial sediment stores to be reconstructed for successive time intervals since deglaciation, and thus for establishing the trajectory of paraglacial landform succession with much greater precision than has hitherto been possible.

Zusammenfassung. Paraglaziale Landformabfolgen spiegeln die unterschiedlichen paraglazialen Subsysteme und deren Entwicklungsstadien über verschiedene Zeitskalen wider. Während manche paraglaziale Sedimentspeicher bereits erodiert werden, unterliegen andere noch der Akkumulation. Über tausende von Jahren sind einzelne Sedimentspeicher (z.B. Schutthalden, Murkegel, Schwemmfächer und Talfüllungen) überwiegend durch Sedimentakkumulation und dann durch Sedimentverlust charakterisiert. Basierend auf der Gleichung $S = (S_a - S_a e^{-\lambda t}) e^{-\kappa t}$ wird ein allgemeines Gleichgewichtsmodell von paraglazialen Sedimentspeichern vorgeschlagen, wobei t = Zeitraum seit Abschmelzen der pleistozänen Gletscher, S = Speichervolumen zu einer Zeit t, S_a = Volumen des verfügbaren Sediments für den Speicheraufbau, λ = Rate des Sedimentaufbrauches von der Sedimentquelle, κ = Rate des Sedimentspeicherabbaus. Es wird gezeigt, dass Sedimentbudgets für jede beliebige Zeit t zurück- oder vorausberechnet werden können durch die Ableitung von Sa, λ und κ. Sie liefert zwei Kurvenwerte, die durch o.g. Gleichung bekannt sind. (t', S_m) und (t", S_v), mit t' = Zeitraum seit Abschmelzen der Gletscher, als der Sedimentspeicher maximales Volumen aufwies, S_m = maximales Sedimentspeichervolumen, t" = verstrichene Zeit seit Abschmelzen der Gletscher, S_v = derzeitig gespeichertes Sedimentvolumen. Neueste Entwicklungen in der Anwendung von geophysikalischen Techniken zur Erfassung der gegenwärtigen (und damit früheren Maxima) Volumina paraglazialer Sedimentspeicher bilden eine wichtige Komponente des vorgestellten Ansatzes. Darin liegt das Potenzial, sich verändernde Sedimentbudgets von individuellen paraglazialen Sedimentspeichern für aufeinanderfolgende Zeitintervalle seit dem Abschmelzen der pleistozänen Gletscher zu rekonstruieren. Damit kann mit viel größerer Genauigkeit eine Zustandskurve paraglazialer Landformabfolgen abgeleitet werden, als dies bisher möglich war.

Introduction

This paper addresses the question of how sediment release, sediment storage and sediment transfer over long ($10^2 - 10^4$ yr) timescales might be calibrated through the study of paraglacial landform succession and sediment storage in mountain valleys that were deglaciated in the Late Pleistocene or Holocene and not subsequently reoccupied by glacier ice. Deglaciated mountain valleys are characterised by a range of distinctive paraglacial sediment stores, such as talus accumulations, avalanche tongues, rockslide or rock avalanche deposits, debris cones, alluvial fans, valley-fill deposits, lacustrine deltas and lake-bottom sediments, that have accumulated at different times and at different rates since deglaciation. By establishing the sequence, timing and magnitude of sediment accumulation, storage and release associated with such landforms, it is feasible to reconstruct the trajectory of postglacial landform change and sediment transfer within deglaciated valley landsystems.

This paper first describes the theoretical basis for paraglacial sediment release, then outlines the concept of paraglacial succession and develops a model of paraglacial sediment storage based on the relative magnitude of sediment inputs and outputs at particular times after deglaciation. Means of calibrating this model by a combination of dating techniques, topographic survey and geophysical profiling are then reviewed.

Paraglacial sediment release: the exhaustion model

Retreating glacier ice frequently exposes landscapes that are in an unstable or metastable state, and consequently liable to sediment release and transfer at rates greatly exceeding background denudation rates. Such glacially-conditioned landscape modification is described as *paraglacial*, a term defined by CHURCH & RYDER (1972) as referring to 'nonglacial processes that are directly conditioned by glaciation', and subsequently broadened to denote 'earth-surface processes, sediments, landforms, landsystems and landscapes directly conditioned by glaciation and deglaciation' (BALLANTYNE 2002a). As all forms of paraglacial activity reflect reworking of nonrenewable sediment sources, BALLANTYNE (2002b) has argued that the rate of sediment release under steady-state conditions can be approximated by the exhaustion model:

$$S_t = S_a e^{-\lambda t} \tag{1}$$

where t is time elapsed since deglaciation, S_t is the proportion of 'available' sediment remaining at time t, S_a is total 'available' sediment at t = 0 and λ is the rate of loss of 'available' sediment by either release or stabilization. If we assign S_a = 1.00 at t = 0, then λ is expressed by:

$$\lambda = \ln(S_t)/ - t. \tag{2}$$

This model implies that rate of sediment release at any time after deglaciation is dependent on the proportion of original sediment available for reworking. The rate of sediment release thus declines exponentially through time (Fig. 1A). Using equation (2), very approximate calibration of the exhaustion model is possible for several primary paraglacial subsystems (Fig. 1B).

Calculation of the changing rate of sediment release on the basis of the volume of sediment in paraglacial sediment stores faces three problems. The first is is that the present volume of many sediment stores represents the difference between sediment input and output since deglaciation, and may therefore under-represent the volume of non-dissolved sediment released since deglaciation. The second is that measurement of sediment volume requires data on sediment depth, a problem considered in the final part of this paper. Finally, most denudation rates calculated on the basis of present volume of paraglacial sediment stores such as talus cones, debris cones, alluvial fans and valley fill are based on retrodiction of *mean* denudation rate (R_d) calculated using some variant of the relation:

$$R_d = V/A\,t \tag{3}$$

where V is the volume of sediment accumulated from a source area A during time (t) since deglaciation. Mean denudation rate (R_d) is related to the exhaustion model by:

$$R_d = \left[\frac{S_a}{t}\int_0^t e^{-\lambda t}dt\right]/A = \left[\frac{S_a}{\lambda t}(1 - e^{-\lambda t})\right]/A. \tag{4}$$

As S_a is usually unknown, equation (4) implies that λ must be determined before the 'stage' in paraglacial sediment release reached can be determined and changes in the rate of sediment release reconstructed. Calculation of λ requires data on denudation rate (or sediment accumulation rate) over at least two different time periods; alternatively, it can be estimated from the time required for sediment exhaustion (BALLANTYNE 2002a).

The behaviour of secondary (primarily fluvial) paraglacial systems in which rates of sediment yield reflect not only release of *in situ* glacigenic sediment and rockwall debris but also reworking of antecedent paraglacial sediment stores may be intrinsically more complex. CHURCH & SLAYMAKER (1989) proposed that reworking of 'upstream' sediment sources

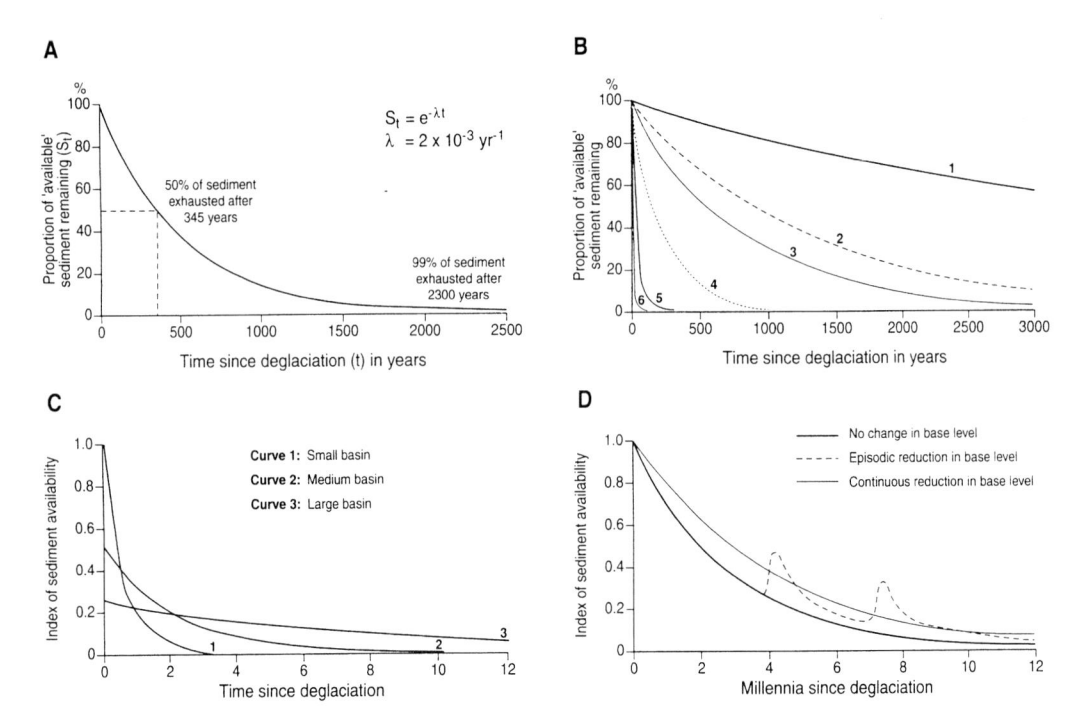

Fig. 1. A. Exhaustion model of paraglacial sediment release. In this example, $\lambda = 0.002$ yr^{-1} (i.e. 0.002% of remaining 'available' sediment is released per year). B. Exhaustion curves for different paraglacial processes: (1) rock-slope failure, from data in CRUDEN & HU (1993); (2) rockfall and talus accumulation, based on data in HINCHLIFFE & BALLANTYNE (1999); (3) accumulation of large alluvial fans (averaged from several sources); (4) rock-slope deformation (averaged from several sources); (5) modification of drift-mantled slopes, averaged from BALLANTYNE (1995) and CURRY (1999); (6) modification of glacier forelands, based on data in MATTHEWS et al. (1998). C. Exhaustion model of paraglacial fluvial sediment transfer in nested basins of different size. The model assumes that initial sediment availability is inversely related to basin size. Rate of sediment reworking declines more gradually with increasing basin size because sediment yield in larger basins is augmented by supply of reworked glacigenic sediment from upstream tributaries. D. An example of extrinsic perturbation on the temporal pattern of paraglacial sediment release: the effect of neotectonically-induced base-level change on sediment flux in fluvial systems. Adapted from BALLANTYNE (2002b).

leads to a lagged peak in paraglacial sediment transport in trunk streams, though the available data (characterized by an increase in specific sediment yield with catchment area) are consistent with the exhaustion model if it is assumed that rates of initial sediment yield are greater in upland tributary basins due to steeper relief (Fig. 1C).

The exhaustion model of paraglacial response assumes steady-state conditions in which there is no change in processes-generating mechanisms or other boundary conditions. Over millennial timescales this assumption may be vitiated. In paraglacial fluvial systems, for example, uplift may cause a fall in base-levels, thus rejuvenating sediment release. Such rejuvenation may be episodic, producing secondary peaks in the rate of sediment transport that reflect temporary increases in the volume of 'available' sediment, or quasi-continuous, effectively slowing the decline in sediment flux by providing new sources of sediment for reworking (Fig. 1D).

Paraglacial succession: concept

The concept of paraglacial succession arises from the fact that different paraglacial subsystems relax over different timescales (Fig. 1B), implying that some depositional landforms (sediment stores) that accumulated soon after deglaciation may experience net erosion and degradation even as other paraglacial sediment stores continue to accumulate. In consequence, the paraglacial landform assemblage evolves towards a 'postglacial climax community' dominated by those features with the greatest formative longevity and persistence in the landscape (Fig. 2).

Evidence for such a paraglacial succession is evident from the work of OWEN et al. (1995), who mapped landform assemblages in three valleys in the Lahul Himalaya that were deglaciated at different times, and quantified the percentage area occupied by different assemblages. Their results show a marked increase through time of the area occupied by alluvial valley fills, slopes modified by mass movement and particularly paraglacial fans, largely at the expense of the area occupied by moraines and thus paraglacially-modified glacier forelands. The area occupied by talus also shows a slight decline, which may represent erosion of talus accumulations formed early in the paraglacial cycle. This study provides insights into paraglacial succession in a mountain environment of unusually high relief, but cannot be used to calibrate or quantify change as the volumes of sediment contained within different landforms is unknown. SCHROTT et al. (2002) have also identified a local paraglacial succession in the Bavarian Alps, in this case dominated initially by talus accumulation, then more recently by floodplain aggradation following damming of the valley by a rockslide 400–600 years ago.

In many mountain areas deglaciated in the Late Pleistocene or Early Holocene, valley-side sediment stores (talus accumulations, debris cones and alluvial fans) are often relict, being extensively vegetated and, in some cases, deeply eroded (e.g. RYDER 1971, BRAZIER et al. 1988, OWEN 1989, BEAUDOIN & KING 1994, MARION et al. 1995, LIAN & HICKIN 1996, OWEN & SHARMA 1998, HINCHLIFFE et al. 1998, HOFFMANN & SCHROTT 2002, SCHROTT et al. 2002, 2003). Erosion of sediment stores implies that net sediment accumulation has given way to net sediment loss. This is also true of terraced paraglacial valley fill deposits, which imply that floodplain aggradation has been succeeded by floodplain incision (JACKSON et al. 1982, RAINS & WELCH 1988). Only in lacustrine depocentres (deltas and lake floor deposits) is a monotonic but declining trend of sediment accumulation recorded in view of the high efficiency of lakes as sediment traps (HINDERER 2001), implying that large lakes can be considered as paraglacial sediment sinks rather than stores, at least until they become completely infilled. A progressive reduction in the rate of sediment storage in paraglacial talus accumulations, debris cones, alluvial fans and valley fills is to be expected from the exhaustion model (equation 1) governing the rate of primary sediment release. Subsequent net *loss* of sediment from storage, however, implies that at some time after deglaciation the rate of sediment output began to exceed the rate of sediment input. As this occurs at different

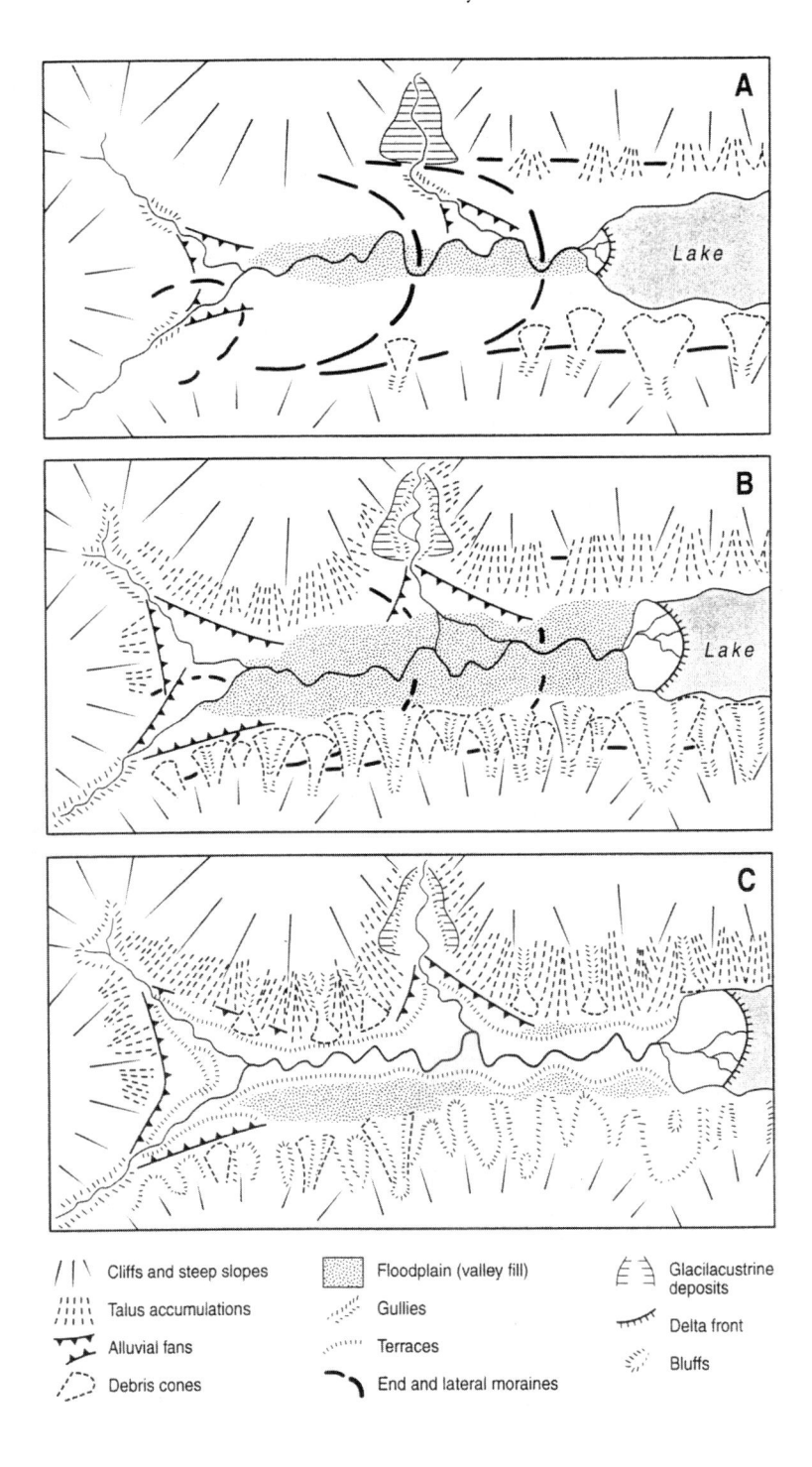

/|\ Cliffs and steep slopes	▒ Floodplain (valley fill)	⩎ Glacilacustrine deposits
///// Talus accumulations	,,,' Gullies	⌒⌒ Delta front
⌵⌵⌵ Alluvial fans	,,,,,,,, Terraces	,,,,, Bluffs
⌒⌒ Debris cones	⌒ End and lateral moraines	

times and at different rates for particular sediment stores (Fig. 1B) then the paraglacial landform assemblage progressively changes through time (Fig. 2).

Paraglacial succession: a general model of sediment storage

For any particular paraglacial sediment store such as a talus cone, debris cone or alluvial fan, the input of nondissolved sediment (S_i) changes at the same rate as sediment loss from the contributory source area, so that at any time (t) after deglaciation:

$$S_i = S_a - S_t = S_a - S_a e^{-\lambda t}. \tag{5}$$

As sediment stores accumulate, however, they are often subject to output losses, for example by debris flow activity, sheetwash, and fluvial erosion on slopes and at the toe of accumulating sediment stores. If we assume that the rate of sediment output is controlled by sediment 'availability' within the sediment store, then the volume of sediment storage (S) at time (t) under steady-state conditions can be expressed as:

$$S = S_i e^{-\kappa t} = (S_a - S_a e^{-\lambda t}) e^{-\kappa t} \tag{6}$$

where κ is the rate of sediment loss from the sediment store. If we set $S_a = 1.00$ for $t = 0$, this simplifies to:

$$S = (1 - e^{-\lambda t}) e^{-\kappa t}. \tag{6a}$$

Sediment output (S_r) at any time (t) after deglaciation is the difference between sediment input (S_i) and sediment storage (S) at that time:

$$S_r = (S_i - S) = (S_a - S_a e^{-\lambda t}) - (S_a - S_a e^{-\lambda t}) e^{-\kappa t} \tag{7}$$

and, if $S_a = 1.00$ at $t = 0$,

$$S_r = (1 - e^{-\lambda t}) - (1 - e^{-\lambda t}) e^{-\kappa t}. \tag{7a}$$

Fig. 2. Paraglacial landform succession (schematic). A. Soon after deglaciation, moraines and glacilacustrine deposits are largely intact; debris cones begin to accumulate on the south size of the valley, talus cones on the north side and alluvial fans at the mouths of tributary valleys; B. After a few millennia many moraines have been eroded or buried and glacilacustrine deposits are deeply eroded; alluvial fans and alluvial valley fill are near their maximum dimensions, though debris cones are already deeply incised; talus continues to accumulate and the delta has prograded into the lake. C. After perhaps 10–15 ka moraines have been largely removed or buried and only fragments of the glacilacustrine deposits remain. Paraglacial debris cones are reduced to eroded bluffs and both alluvial fans and the floodplain have been entrenched and terraced. Erosion of some talus accumulations has commenced, with concomitant accumulation of a second generation of debris cones. Only the delta has a net positive sediment budget, and continues to prograde into the lake.

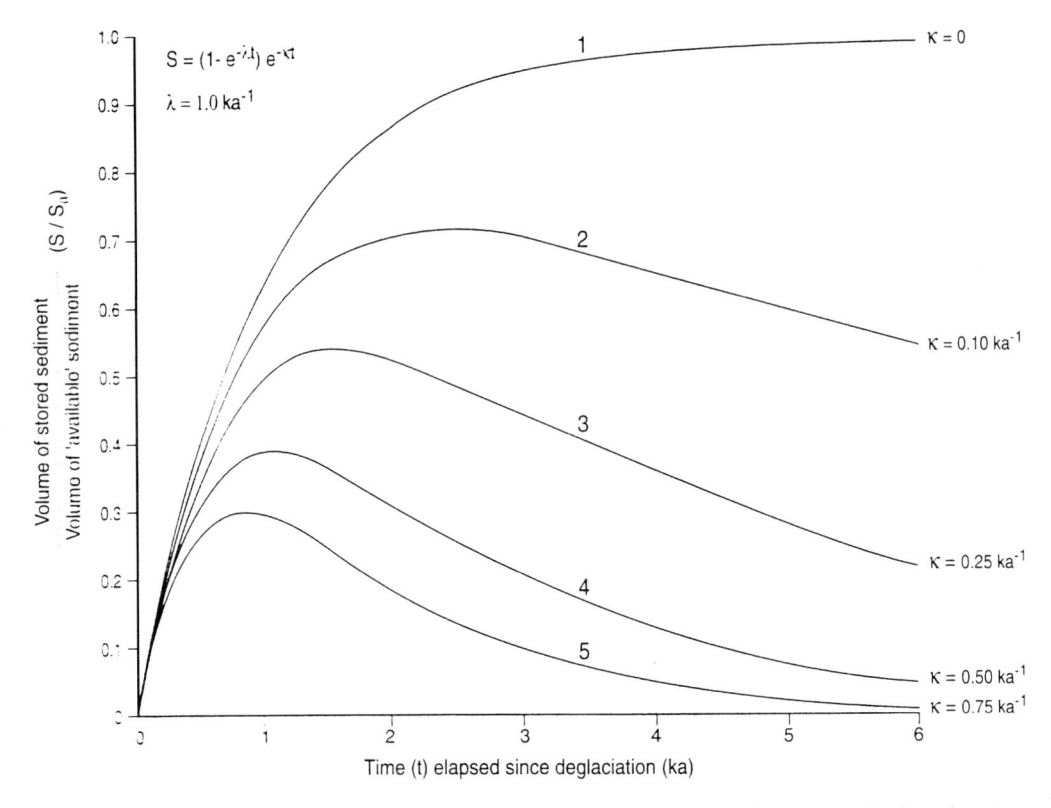

Fig. 3. Change in volume of sediment storage as a fraction of 'available' sediment (S/Sa) plotted against time elapsed since deglaciation for $\lambda = 1.0$ ka^{-1} and κ = zero, 0.1 ka^{-1}, 0.25 ka^{-1}, 0.5 ka^{-1} and 0.75 ka^{-1}. For explanation, see text.

Equation (6a) can be used to generate a series of curves that describe the pattern of sediment accumulation through time for individual sediment stores. In Fig. 3, λ is given a value of 1.0 ka^{-1}, and κ is assigned values of zero, 0.10 ka^{-1}, 0.25 ka^{-1}, 0.50 ka^{-1} and 0.75 ka^{-1} to illustrate the effect of differing rates of sediment output on the trajectory of sediment storage. When $\kappa = 0$, output is zero and thus at any time (t) since deglaciation the volume of sediment storage (S) is equal to cumulative sediment input (S_i), and thus cumulative sediment loss (S_r) from the sediment source (curve 1, Fig. 3). Curves 2–5 show that in all cases where $\kappa > 0$, rate of sediment accumulation slows then becomes negative, implying that the change from net sediment accumulation to net sediment loss is an *intrinsic* function of paraglacial sediment stores under steady-state conditions, and not necessarily due to changing boundary conditions.

The size of the difference between λ and κ determines both the maximum volume of sediment storage and when maximum storage occurs (t′). When κ/λ is large, the maximum volume stored is much less than the volume of sediment available for reworking (S_a) and maximum storage is achieved rapidly. For curve 5 in Fig. 3, $\kappa/\lambda = 0.75$, the maximum volume stored is 0.302 S_a, and t′ occurs 845 years after deglaciation. Conversely, when κ/λ is small, a much larger proportion of sediment enters storage, and maximum volume of stored sediment occurs later: for curve 2, $\kappa/\lambda = 0.1$, the maximum volume stored is 0.715 S_a, and t' occurs about 2.3 ka after deglaciation. The maximum volume of sediment stored [max (S)] and the time elapsed since deglaciation when maximum storage occurs (t') can be derived from equation (6a) as:

$$\max(S) = [\lambda/(\lambda + \kappa)] \cdot [\kappa/(\lambda + \kappa)]^{\kappa/\lambda} \qquad (8)$$

and

$$t' = -\ln[\kappa/(\lambda + \kappa)]/\lambda. \qquad (9)$$

These equations imply that [max (S)] is a simple function of κ/λ, irrespective of the actual values involved (Fig. 4A), whereas the relationship between t', λ and κ is more complex, being dependent on the actual values of λ and κ (Figure 4B). Simultaneous solution of equations (8) and (9) can be used to calculate the values of λ and κ for individual paraglacial sediment stores, but only when (1) [max (S)] can be expressed as a fraction of S_a, rather than as an absolute volume, and (2) when t' is known. For example, data presented in BALLANTYNE (1995) for a small paraglacial debris cone that formed since recent deglaciation in Bergsetdalen (Norway) yields best-estimate values of [max (S)] = 0.90 S_a and t' = 162 yr; solving for λ and κ by simultaneous solution of equations (8) and (9) yields $\lambda = 23.755$ ka^{-1} and $\kappa = 0.517$ ka^{-1}. Substitution of these values into equation (6a) implies that this sediment store will be reduced to about 54% of its maximum volume at t = 1.0 ka, 19% at t = 3.0 ka and 7% at t = 5.0 ka. This example reflects unusually rapid accumulation and subsequent slower degradation of a paraglacial sediment store. For a large alluvial fan for which [max (S)] = 0.5 (i.e. 0.5 S_a) and t' = 5 ka, the values of λ and κ obtained by substitution into equations (8) and (9) are 0.296 ka^{-1} and 0.087 ka^{-1} respectively. Substitution of these values into equation (6a) indicates that S will be reduced to 0.4 (i.e. 80% of its maximum volume) approximately 10 ka after deglaciation.

As the value of S_a is often unknown, however, a more general procedure for calculating λ and κ is required. This can be achieved if four values are known: (1) the time elapsed since deglaciation at which a sediment store achieved its maximum volume (t'); (2) the time interval between deglaciation and the present (t''); (3) the maximum volume of the sediment store (S_m) at t'; and (4) the present volume of stored sediment (S_v). Note that the terms S_m and S_v refer to *measured* volumes (in m³) rather than storage volume as a proportion of S_a. As the points (t', S_m) and (t'', S_v) both lie on the curve described by equation (6), then:

$$S_m = (S_a - S_a e^{-\lambda t'}) e^{-\kappa t'} \qquad (10)$$

and

$$S_v = (S_a - S_a e^{-\lambda t''}) e^{-\kappa t''}. \qquad (11)$$

As S = 0 when t = 0, three points on the curve defined by equation (6) are known: the origin (t_0, S_0), the maximum value of S (t', S_m) and the present value of S (t'', S_v) (Fig. 5). Although it is not feasible to derive general mathematical expressions for S_a, λ and κ from these limiting points, it is nonetheless possible to solve for all three parameters when the numerical values of

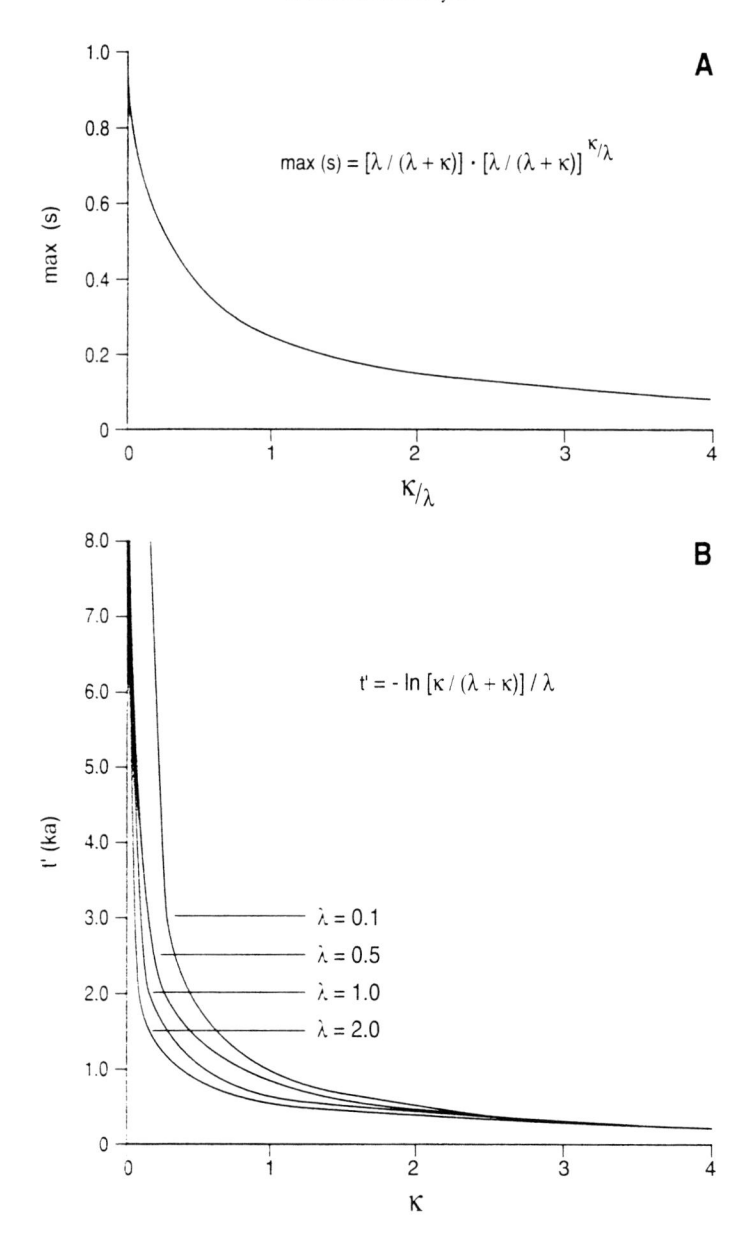

Fig. 4. A. Relationship between the maximum volume of sediment entering storage and the ratio κ/λ. B. Relationship between the time (t') at which maximum sediment storage occurs and the value of κ for $\lambda = 0.1, 0.5, 1.0$ and 2.0. For explanation, see text.

t', t'', S_m and S_v are known. As t' is the same for both [max (S)] and S_m, equation 8 can be rewritten:

$$S_m = S_a \, [\lambda/(\lambda + \kappa)] \cdot [\kappa/(\lambda + \kappa)]^{\kappa/\lambda}. \tag{12}$$

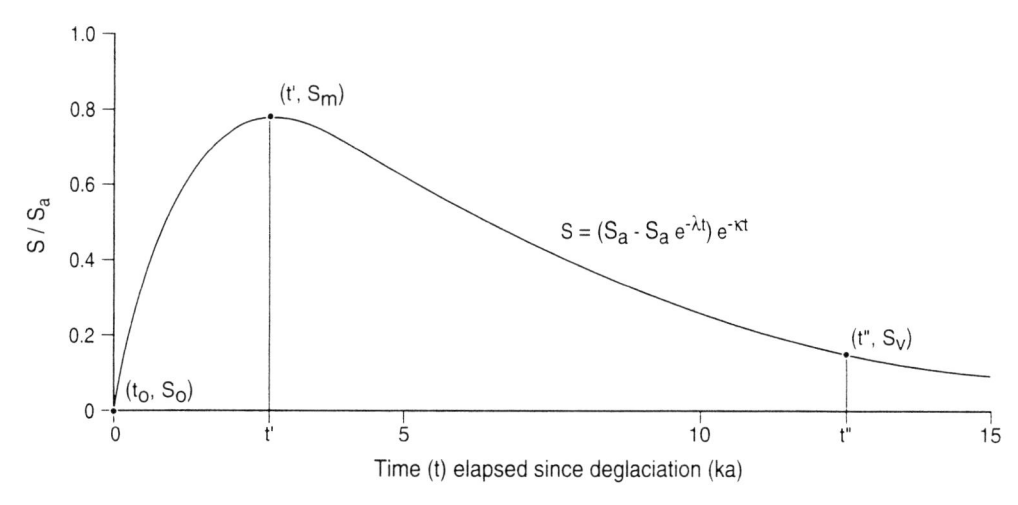

Fig. 5. Curve showing the changing volume of sediment storage described by the equation $S = (S_a - S_a e^{-\lambda t}) e^{-\kappa t}$, showing the points on the curve that are known or can be determined empirically: (t_0, S_0), (t', S_m) and (t'', S_v). For explanation, see text.

For any (known) value of S_m it is therefore possible to generate paired solutions for λ and κ for unit increments of S_a. The equivalent values of S_a, λ and k thus obtained are then entered iteratively into equation (11) until a solution is obtained that corresponds to the (known) values of S_v and t''. Once a solution has been achieved for S_a, λ and κ, it is possible to quantify all parameters of the model (S_t, S_i, S and S_r) for any given time (t) since deglaciation (equations 1, 5, 6 and 7). Mean denudation rate (R_d) at any time (t) can also be calculated, providing that the area (A) of the sediment source is known (equation 4).

The sediment storage model: a hypothetical example

Given the input parameters $S_m = 500{,}000$ m³, $S_v = 400{,}000$ m³, $t' = 5$ ka and $t'' = 12$ ka for a particular paraglacial store such as a talus accumulation, debris cone or alluvial fan, the above procedure yields $S_a = 672940$ m³, $\lambda = 0.5102$ ka^{-1} and $\kappa = 0.0432$ ka^{-1}. From these values it is possible to retrodict or predict the sediment budget associated with the sediment store at any time (t) since deglaciation. After 2 ka, for example, of the initial 672,940 m³ of sediment available for reworking at deglaciation (S_a), the volume still available (S_t) is 242,567 m³, the cumulative input to the sediment store (S_i) amounts to 430,373 m³, the volume of sediment stored (S) is 394,777 m³ and the cumulative output from the sediment store is 11,361 m³. After 10 ka, the equivalent values are $S_t = 4095$ m³, $S_i = 668{,}845$ m³, $S = 434{,}368$ m³ and $S_r = 234{,}477$ m³. At any time (t) the relationship between these parameters conforms to the general sediment balance equation:

$$S_a - S_t = S_i = S + S_r. \tag{13}$$

If the area of the sediment source (A) is known, mean denudation rate (R_d) can be calculated from equation (4). For example, if A = 100,000 m² for the above inputs, R_d at t = 2 ka is 4.21 m ka⁻¹ (= 4.21 mm yr⁻¹) and R_d at t = 10 ka is 1.31 m ka⁻¹ (= 1.31 mm yr⁻¹). Note that R_d refers to the mean denudation rate over the time period (t − t₀), not the rate of denudation occurring at time t. By plotting the values of S_t, S_i, S and S_r over successive time intervals (Fig. 6) the changing sediment budget of a particular paraglacial sediment store can be traced, allowing reconstruction of the associated sediment flux. If this can be achieved for a range of sediment stores (talus sheets and cones, debris cones, alluvial fans, valley fills, lake deltas and lake-floor sediments; Fig. 2) then the nature of paraglacial succession in deglaciated mountain valleys can be reconstructed and the changing sediment budget for the valley as a whole can be estimated.

Calibrating the model: (1) dating

Calibration of the above model requires data on four variables, two of which reflect time since deglaciation (t′ and t″) and two of which reflect the volume of sediment stored in a particular

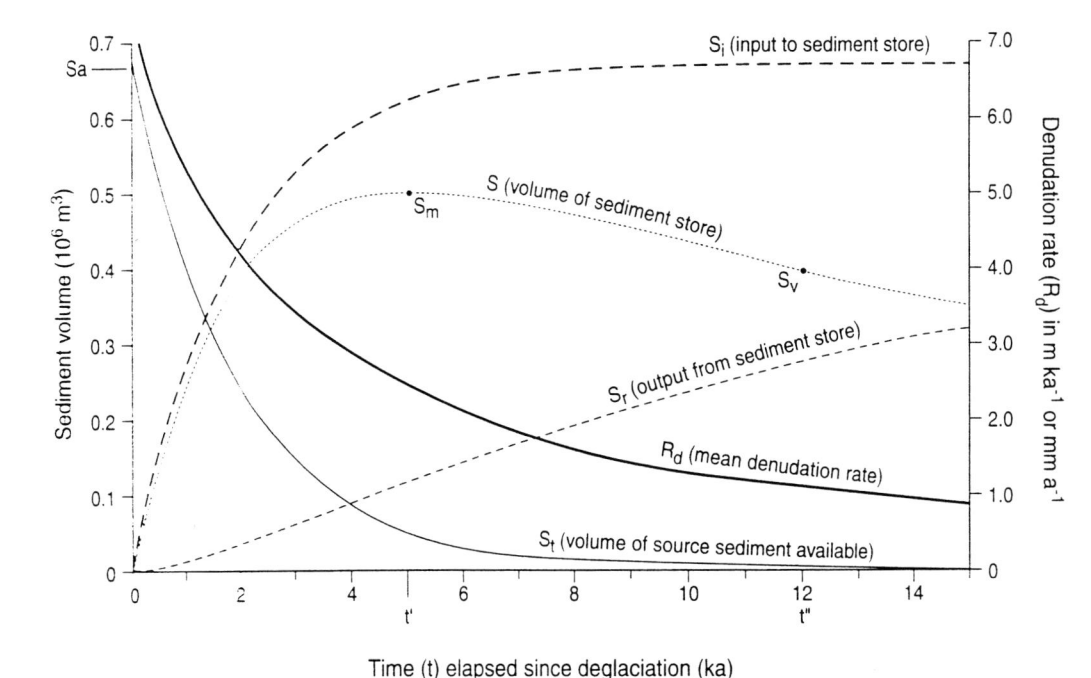

Fig. 6. Changing sediment budget of a paraglacial sediment store for which S_m = 500,000 m³ at t′ = 5 ka and S_v = 400,000 m³ at t″ = 12 ka. S_i, S, S_r and S_t are measured on the left hand scale, and changing mean denudation rate (R_d) on the right hand scale. For explanation, see text.

depositional landform (S_m and S_v) at these times. Time elapsed between deglaciation and the present (t'') is often known, at least approximately, from reconstructions of local glacial history. In studies of paraglacial sediment accumulation since Late Pleistocene or Early Holocene deglaciation, the age of deglaciation is usually underpinned by radiocarbon dating (e.g. MILLER et al. 1993, FRIELE et al. 1999, MÜLLER 1999), though cosmogenic isotope dating is now increasingly used to determine deglaciation age (GOSSE & PHILLIPS 2001). In studies of recent paraglacial sediment flux, timing of deglaciation is usually established from historical evidence or lichenometry (e.g. BALLANTYNE & BENN 1994, BALLANTYNE 1995).

Establishing the time (t') at which the sediment storage reached a maximum has been achieved by tephrochronology (e.g. RYDER 1971, EYLES et al. 1988, BEAUDOIN & KING 1994), radiocarbon dating (e.g. JACKSON et al. 1982, LIAN & HICKIN 1996, WATANABE et al. 1998, FRIELE et al. 1999, BALLANTYNE & WHITTINGTON 1999), stratigraphigraphic constraints (PEACOCK 1986), the timing of tree colonisation of sediment stores (JACKSON et al. 1982, MARION et al. 1995) and, in the case of recent paraglacial sediment accumulations, dendrochronology and lichenometry (BALLANTYNE 1995, HARRISON & WINCHESTER 1997). For accumulations of coarse paraglacial sediments, such as rockslide deposits and large boulders on talus, cosmogenic isotope dating offers a means of determining when rock-slope failure occurred or talus accumulation reached its maximum (BALLANTYNE et al. 1998).

Calibrating the model: (2) the role of geophysical profiling

The volumetric parameters S_v and S_m are derived by a twofold procedure, involving (1) evaluating the present volume stored (S_v) and (2) evaluating the net volume of sediment loss from a particular sediment store after it reached its maximum volume. The two figures are then added to yield the value of S_m, the maximum volume of sediment storage.

Evaluation of the net volume of sediment loss from a paraglacial sediment store that has reached its maximum volume then experienced subsequent incision has been accomplished by topographic survey, photogrammetry and use of digital elevation models (e.g. CAMPY et al. 1998, WATANABE et al. 1998, BALLANTYNE & WHITTINGTON 1999, CURRY 1999) The general approach has been to equate sediment loss with the volume of gullies or valleys cut into sediment stores. Assessment of the present volume of paraglacial sediment stores is often more problematic, however, due to the difficulty of establishing the depth of sediment accumulation. In rare cases, gully incision to bedrock provides data to constrain sediment depth (HINCHLIFFE & BALLANTYNE 1999). More commonly, depth to rockhead has been based on assumptions regarding the configuration of the underlying bedrock (e.g. BALLANTYNE 1995, WATANABE et al. 1998, SHRODER et al. 1999). Such assumptions, however, may lead to overestimation of sediment volume (HOFFMANN & SCHROTT 2002, SCHROTT & ADAMS 2002, SCHROTT et al. 2003).

Recent research, however, has confirmed the feasibility of employing geophysical techniques to establish the depth profile (and thus the volume) of paraglacial sediment stores. SASS & WOLNY (2001) employed ground penetrating radar (GPR) to establish the depths of paraglacial talus accumulations. This system delivered subsurface data to 70 m depth, and showed that the talus/bedrock contact occurred at depths of 5–25 m, though it was not detected continuously in all profiles. HOFFMAN & SCHROTT (2002) have demonstrated the successful application of refraction seismics to determination of talus thickness. They detected underlying bedrock at depths of 3–24 m with an estimated accuracy of ± 20%, but noted a potential drawback in that the technique may not distinguish between talus debris and underlying till, thus potentially overestimating the former if the latter is present.

Estimation of the thickness of sediments in paraglacial alluvial fans, debris cones and alluvial fills has also been shown to be amenable to geophysical profiling. Using a cross-section of sediment configuration constructed from GPR data, FRIELE et al. (1999) calculated the volumes of sediment deposited on a large fan in British Columbia during the time intervals 10.2–6.0 ^{14}C ka BP and 6.0 ^{14}C ka BP to the present, and concluded that approximately 90% of sediment accumulated in the former interval. In northern Sweden, BAUMGART-KOTARBA et al. (2001) used 1D resistivity techniques to model the depth and configuration of postglacial valley fill. Using a combination of seismic reflection and borehole data, MÜLLER (1999) reconstructed the accumulation of deltaic, inter-deltaic and fan deposits since deglaciation in a thick fill in eastern Switzerland, and charted the progressive slowing of paraglacial sediment inputs through time.

In a glaciated valley in the Dolomites, SCHROTT & ADAMS (2002) employed seismic refraction and geoelectrical resistivity sounding to generate a digital elevation model (DEM) of subsediment bedrock topography. The detection limit of the seismic techniques they employed was c. 50 m. They calculated the volume of accumulated sediment by subtracting the rockhead DEM from a ground-surface DEM, and showed that talus, debris cones and avalanche cones represent almost 70% of total paraglacial sediment storage. A detailed inventory of paraglacial sediment storage has also been carried out by SCHROTT et al. (2002, 2003) in the Reintal, a glacial trough in the Bavarian Alps. The volumes of sediment stores were calculated as the difference between ground-surface and rockhead DEMs based on 66 seismic refraction profiles. This study demonstrated that at this high-relief site talus accumulations dominate sediment storage, followed by rockslide (bergsturtz) deposits, alluvial fans and floodplain deposits, debris and avalanche cones.

The coarse deposits emplaced on valley floors by major rock-slope failures also appear amenable to geophysical volume estimation. In the Reintal, this was achieved by seismic refraction profiling (SCHROTT et al. 2003). Elsewhere, a range of geophysical techniques including GPR soundings, DC resistivity, seismic refraction profiling and gravimetry have been employed to reconstruct the structure and depth of rock glaciers (BARSCH 1996, BURGER et al. 1999). Such rock glaciers resemble landslide runout deposits in terms of the depth and thickness of constituent debris, suggesting that such methods may be equally applicable for establishing the depth and volume of paraglacial landslide deposits.

In many mountain areas, paraglacial sediment transport terminates downvalley in glacially-overdeepened lake basins. Because such basins are effective sediment traps, they represent sinks in which sediment continues to accumulate even after other paraglacial sediment stores have crossed the threshold from net accumulation to net erosion. For this reason, evaluation of the volume of sediment that has accumulated in lake basins since deglaciation is fundamental for reconstructing the changing sediment budget of paraglacial landform successions in many glaciated mountain valleys. In distal lake basins, paraglacial sedimentation is characterised by delta progradation and the accumulation of lake bottom deposits. Although a rough indication of sediment delivery to lakes may be inferred from the rate of delta advance (JORDAN & SLAYMAKER 1991), geophysical methods offer a more accurate means of assessing total sediment infill. In the fjord lakes of British Columbia, this has been achieved by seismic reflection survey and accoustic profiling, which have been employed to determine the depth and volume of postglacial sediment units overlying glacilacustrine deposits (EYLES et al. 1990, MULLINS et al. 1990, DESLOGES & GILBERT 1991, GILBERT & DESLOGES 1992). HINDERER (2001) provides a review of the role of seismic profiling, often in combination with geomorphological evidence and coring data, in determining the volume of postglacial sediment accumulation in 16 major valleys and lake basins in the Alps. His survey highlights rapid sediment accumulation during and immediately after deglaciation, with a progressive reduction in sediment influx during much of the Holocene.

The above review indicates the potential of geophysical profiling for determining the volumes of paraglacial sediment stores, and thus for reconstructing the changing budget of paraglacial sediment flux in deglaciated mountain valleys. The use of geophysical techniques in this way nonetheless has two potential limitations. In high-energy paraglacial environments, sediment depth may exceed the capability of geophysical sensing. In the Karakoram Mountains, for example, valley fills exceed 100 m in depth (OWEN 1989), and may pose problems of penetration to rockhead. More generally, differences in subsurface sediment characteristics may be difficult to detect, so that use of the sediment/rockhead boundary to assess sediment depth and volume may overestimate paraglacial sediment storage by including an indeterminate volume of underlying glacigenic sediment. The best circumstances for accurate reconstruction of the volume of paraglacial sediment stores would appear to be where sediment cover is relatively thin (< 50 m) and where geophysical measurements are supplemented by borehole data and take account of geomorphometric constraints (SCHROTT et al. 2002, 2003).

Conclusions

Despite a recent upsurge in field research on paraglacial processes, landforms and sediment flux, development of the theoretical basis of paraglacial landscape modification has been limited. The concept of paraglacial landform succession and the model of paraglacial sediment storage outlined here represent extensions of earlier work (BALLANTYNE 2002a, 2002b) that attempted to identify the unifying characteristics of the paraglacial concept. The sediment storage model developed above (equation 6) implies three axioms governing the behaviour of paraglacial sediment stores, namely:

(1) that an eventual change from net (nondissolved) sediment accumulation to net (nondissolved) sediment loss is an *intrinsic* characteristic of all paraglacial sediment stores unless sediment loss is zero, even under steady-state conditions;
(2) that the maximum volume of sediment accumulation can be related to the ratio between rate of sediment loss (κ) and the rate of sediment accumulation (λ); where κ/λ is low, a large proportion of 'available' sediment may be stored, and *vice-versa*;

(3) that the timing of maximum sediment storage is also determined by the rates of sediment accumulation (λ) and loss (κ); for any given value of λ, the time elapsed since deglaciation at which maximum storage occurs is an inverse nonlinear function of κ/λ.

Theoretical modelling of sediment inputs and outputs associated with paraglacial sediment stores that are fed from a discrete volume of 'available' sediment demonstrates the feasibility of reconstructing the changing sediment budget of individual sediment stores on the basis of four parameters, namely time elapsed since deglaciation, the present volume of stored sediment, the maximum volume of stored sediment prior to the onset of net erosion, and the time when maximum volume was achieved. Such modelling assumes steady-state conditions. This assumption is probably reasonable for slope-foot sediment stores that accumulate independently of extrinsic base-level changes. Fluvially-dominated systems such as alluvial valley fills and alluvial fans are probably more vulnerable to base-level change induced by neotectonic uplift or tilting over millennial timescales, and the applicability of the model to alluvial sediment stores will therefore depend on local circumstances.

Critical to the calibration of the sediment storage model is accurate determination of the volume of stored sediment. Recent advances in the application of geophysical approaches, summarised above, have demonstrated the feasibility of employing GPR, seismic refraction, seismic reflection, resistivity and other techniques to determine the depth of the sediment/rockhead boundary, and thereby to permit calculate the volume of stored sediment.

Most studies of paraglacial landscape change in mountain environments have been essentially static, or based on temporal 'snapshots' of changing paraglacial landforms (e.g. CURRY 1999) or landform associations (e.g. OWEN et al. 1995, SCHROTT et al. 2002). Both theoretically and technologically, however, it now appears feasible to develop a dynamic approach to the study of paraglacial landform succession in deglaciated mountain valleys, thus opening up a new and exciting approach to understanding the way in which mountain landscapes have responded to glacial inheritance.

Acknowledgements

The author thanks Professor Edmund Robertson and Dr. Ruth Robinson for mathematical assistance, Professors Nel Caine and Colin Thorn for comments, and Graeme Sandeman for preparing the figures.

References

BALLANTYNE, C. K. (1995): Paraglacial debris cone formation on recently-deglaciated terrain. – The Holocene **5**: 25–33.
BALLANTYNE, C.K: (2002a): Paraglacial geomorphology. – Quatern. Sci. Rev. **21**: 1935–2017.
BALLANTYNE, C.K: (2002b): A general model of paraglacial landscape response. – The Holocene **12**: 371–376.
BALLANTYNE, C. K. & BENN, D. I. (1994): Paraglacial slope adjustment and resedimentation following glacier retreat, Fåbergstølsdalen, Norway. – Arct. Alpine Res. **26**: 255–269.
BALLANTYNE, C. K., STONE, J. O. & FIFIELD, L. K. (1998): Cosmogenic Cl-36 dating of postglacial landsliding at The Storr, Isle of Skye, Scotland. – The Holocene **8**: 347–351.

BALLANTYNE, C. K. & WHITTINGTON, G. (1999): Late Holocene floodplain incision and alluvial fan formation in the central Grampian Highlands, Scotland: chronology, environment and implications. – Journ. Quatern. Sci. **14**: 651–671.

BARSCH, D. (1996): Rockglaciers: indicators for the present and former geoecology in high mountain environment. – 331 pp., Springer-Verlag, Berlin.

BAUMGART-KOTARBA, M., KEDZIA, S., KOTARBA, A. & MOSCICKI, J. (2001): Geomorphological and geophysical studies in a subarctic environment of Kärkevagge Valley, Abisko Mountains, Northern Sweden. – Bull. Polish Acad. Scis.: Earth Scis. **49**: 123–135.

BEAUDOIN, A. B. & KING, R. H. (1994): Holocene palaeoenvironmental record preserved in a paraglacial alluvial fan, Sunwapta Pass, Jasper National Park, Alberta, Canada. – Catena **22**: 227–248.

BRAZIER, V., WHITTINGTON, G. & BALLANTYNE, C. K. (1988): Holocene debris cone evolution in Glen Etive, Western Grampian Highlands, Scotland. – Earth Surf. Process. Landforms **13**: 525–531.

BURGER, K. C., DEGENHARDT, J. J. & GIARDINO, J. R. (1999): Engineering geomorphology of rock glaciers. – Geomorphology **31**: 93–132.

CAMPY, M., BUONOCRISTIANI, J. F. & BICHET, V. (1998): Sediment yield from glacio-lacustrine calcareous deposits during the postglacial period in the Combe d'Ain (Jura, France). – Earth Surf. Process. Landforms **23**: 429–444.

CHURCH, M. & RYDER, J. M. (1972): Paraglacial sedimentation: a consideration of fluvial processes conditioned by glaciation. – Geol. Soc. Amer. Bull. **83**: 3059–3071.

CHURCH, M. & SLAYMAKER, O. (1989): Disequilibrium of Holocene sediment yield in glaciated British Columbia. – Nature **337**: 452–454.

CRUDER, D. & HU, X. Q. (1993): Exhaustion and steady-state models for predicting landslide hazards in the Canadian Rocky Mountains. – Geomorphology **8**: 279–285.

CURRY, A. M. (1999): Paraglacial modification of slope form. – Earth Surf. Process. Landforms **24**: 1213–1228.

DESLOGES, J. R. & GILBERT, R. E. (1991): Sedimentary record of Harrison Lake: implications for deglaciation in southwestern British Columbia. – Canad. Journ. Earth Scis. **28**: 800–815.

EYLES, N., EYLES, C. H. & MCCABE, A. M. (1988): Late Pleistocene subaerial debris-flow facies of the Bow Valley, near Banff, Canadian Rocky Mountains. – Sedimentology **35**: 465–480.

EYLES, N., MULLINS, H. T. & HINE, A. C. (1990): Thick and fast: sedimentation in a Pleistocene fiord lake of British Columbia, Canada. – Geology **18**: 1153–1157.

FRIELE, P. A., EKES, C. & HICKEN, E. J. (1999): Evolution of Cheekye fan, Squamish, British Columbia: Holocene sedimentation and implications for hazard assessment. – Canad. Journ. Earth Scis. **36**: 2023–2031.

GILBERT, R. & DESLOGES, J. R. (1992): The late Quaternary sedimentary record of Stave Lake, southwestern British Columbia. – Canad. Journ. Earth Scis. **29**: 1997–2006.

GOSSE, J. C. & PHILLIPS, F. M. (2001): Terrestrial in situ cosmogenic nuclides: theory and application. – Quatern. Sci. Rev. **20**: 1475–1560.

HARRISON, S. & WINCHESTER, V. (1997): Age and nature of paraglacial debris cones along the margins of the San Rafael glacier, Chilean Patagonia. – The Holocene **7**: 481–487.

HINCHLIFFE, S. & BALLANTYNE, C. K. (1999): Talus accumulation and rockwall retreat, Trotternish, Isle of Skye, Scotland. – Scott. Geograph. Journ. **115**: 53–70.

HINCHLIFFE, S., BALLANTYNE, C. K. & WALDEN, J. (1998): The structure and sedimentology of relict talus, Trotternish, northern Skye, Scotland. – Earth Surf. Process. Landforms **23**: 545–560.

HINDERER, M. (2001): Late Quaternary denudation of the Alps, valley and lake fillings and modern river loads. – Geodinamica Acta **13**: 1–33.

HOFFMANN, T. & SCHROTT, L. (2002): Modelling sediment thicknesses and rockwall retreat in an Alpine valley using 2D-seismic refraction (Reintal, Bavarian Alps). – Z. Geomorph. N. F., in press.

HOOKE, R. LeB. (2000): Toward a uniform theory of clastic sediment yield in fluvial systems. – Geol. Soc. Amer. Bull. **112**: 1778–1786.

JACKSON, L. E., MACDONALD, G. M. & WILSON, M. C. (1982): Paraglacial origin for terraced river sediments in Bow Valley, Alberta. – Canad. Journ. Earth Scis. **19**: 2219–2231.

Jordan, P. & Slaymaker, O. (1991): Holocene sediment production in Lillooet River, British Columbia: a sediment budget approach. – Géogr. Phys. Quatern. **45**: 45–57.

Lian, O. B. & Hickin, E. J. (1996): Early postglacial sedimentation of lower Seymour Valley, southwestern British Columbia. – Géogr. Phys. Quatern. **50**: 95–102.

Marion, J., Filion, L. & Hétu, B. (1995): The Holocene development of a debris slope in subarctic Québec, Canada. – The Holocene **5**: 409–419.

Matthews, J. A., Shakesby, R. A., Berrisford, M. S. & McEwen, L. J. (1998): Periglacial patterned ground in the Styggedalsbreen glacier foreland, Jotunheimen, southern Norway: micro-topographical, paraglacial and geochronological controls. – Permafrost Periglac. Process. **9**: 147–166.

Miller, D. C., Birkeland, P. W. & Rodbell, D. T. (1993): Evidence for Holocene stability of steep slopes, northern Peruvian Andes, based on soils and radiocarbon dates. – Catena **20**: 1–12.

Müller, B. U. (1999): Paraglacial sedimentation and denudation processes in an Alpine valley of Switzerland. An approach to the quantification of sediment budgets. – Geodinamica Acta **12**: 291–301.

Mullins, H. T., Eyles, N. & Hinchley, E. J. (1990): Seismic reflection investigation of Kalamalka Lake: a 'fiord lake' on the interior plateau of southern British Columbia. – Canad. Journ. Earth Scis. **27**: 1225–1235.

Owen, L. A. (1989): Terraces, uplift and climate in the Karakoram Mountains, northern Pakistan: Karakoram intermontane basin evolution. – Z. Geomorph. N. F., Suppl.-Bd. **76**: 117–146.

Owen, L. A., Benn, D. I., Derbyshire, E., Evans, D. J. A., Thompson, D., Richarddson, S., Lloyd, M. & Holden, C. (1995): The geomorphology and landscape evolution of the Lahul Himalaya, Northern India. – Z. Geomorph. N. F. **39**: 145–174.

Owen, L. A. & Sharma, M. C. (1998): Rates and magnitudes of paraglacial fan formation in the Garwhal Himalaya: implications for landscape evolution. – Geomorphology **26**: 171–184.

Peacock, J. D. (1986): Alluvial fans and an outwash fan in upper Glen Roy, Lochaber. – Scott. Journ. Geol. **22**: 347–366.

Rains, B. & Welch, J. (1988): Out-of-phase Holocene terraces in part of the North Saskatchewan River basin, Alberta. – Canad. Journ. Earth Scis. **25**: 454–464.

Ryder, J. M. (1971): The stratigraphy and morphology of para-glacial alluvial fans in south-central British Columbia. – Canad. Journ. Earth Scis. **8**: 279–298.

Sass, O. & Wolny, K. (2001): Investigations regarding alpine talus slopes using ground-penetrating radar (GPR) in the Bavarian Alps, Germany. – Earth Surf. Process. Landforms **26**: 1071–1086.

Schrott, L. & Adams, T. (2002): Quantifying sediment storage and Holocene denudation in an Alpine basin, Dolomites, Italy. – Z. Geomorph. N. F., Suppl.-Bd. **128**: 129–145.

Schrott, L., Niederheide, A., Hankammer, M., Hufschmidt, G. & Dikau, R. (2002): Sediment storage in a mountain catchment: geomorphic coupling and temporal variability (Reintal, Bavarian Alps, Germany). – Z. Geomorph., Suppl.-Bd. **127**: 175–196.

Schrott, L., Hufschmidt, G., Hankammer, M., Hoffmann, T. & Dikau, R. (2003): Spatial distribution of sediment storage types and quantification of valley fill deposits in an Alpine basin, Reintal, Bavarian Alps, Germany. – Geomorphology, **55**, in press.

Shroder, J. F., Scheppy, R. A. & Bishop, M. P. (1999): Denudation of small alpine basins, Nanga Parbat Himalaya, Pakistan. – Arct., Antarct. Alpine Res. **31**: 121–127.

Watanabe, T., Dali, L. & Shiraiwa, T. (1998): Slope denudation and supply of debris to cones in Landtang Himal, Central Nepal Himalaya. – Geomorphology **26**: 185–197.

Address of the author: Professor Colin K. Ballantyne, School of Geography and Geosciences, University of St Andrews, Fife KY16 9AL, Scotland, UK.

One-dimensional DC-resistivity depth soundings as a tool in permafrost investigations in high mountain areas of Southern Norway

Bernd Etzelmüller, Ivar Berthling, Oslo, and Rune S. Ødegård, Gjøvik, Norway

with 8 figures and 1 table

Summary. The phase change of water from an unfrozen to a frozen state in soils and rocks reduces the ability to conduct electricity, offering the possibility to identify frozen ground and to determine its characteristics such as type of ice by using geophysical measurements. DC resistivity sounding is useful in this respect, and gives indications of permafrost thickness. This paper summarises 1D resistivity measurements on blocky surfaces in or near the permafrost zone in southern Norway. The objective of the paper is to assess the feasibility of the method in permafrost mapping and model validation. The paper shows interpretation from various sites in southern Norway, ranging from bouldery block fields via sites with low sediment cover to non-permafrost sites dominated by coarse ground moraines.

Zusammenfassung. Der Phasenübergang vom ungefrorenen zum gefrorenen Zustand von Wasser in Böden und Festgesteinen reduziert die elektrische Leitfähigkeit des Materials und eröffnet somit die Möglichkeit, gefrorenen Untergrund zu identifizieren und dessen Eigenschaften mit Hilfe geophysikalischer Sondierungen zu erfassen. Geoelektrische Sondierungen sind ein potentiell hilfreiches Werkzeug in dieser Hinsicht, da die Methode Indikationen zur Permafrostmächtigkeit liefert und unterschiedliche genetische Eistypen unterscheiden kann. Dieser Aufsatz fasst eindimensionale geoelektrische Messungen innerhalb der Permafrostzone auf blockreichen Oberflächen im südlichen Norwegen zusammen. Das Ziel der Arbeit ist es, die Methode mit Hinsicht auf deren Anwendung zur Permafrostkartierung und Modellvalidierung zu diskutieren. Interpretationen von verschiedenen Standorten im südlichen Norwegen, von Blockfeldern über Standorte mit geringer Sedimentmächtigkeit bis permafrostfreien, von grobkörniger Grundmoräne dominierten Lokalitäten werden veranschaulicht. Der Aufsatz zeigt die Variabilität der Resultate und deren mögliche Interpretation.

1 Introduction and background

Mountain permafrost occurs usually sporadic and discontinuous over larger areas. In alpine environments, the discontinuity of permafrost is especially caused by topography, as valleys normally dissect high-altitude areas, but also by strong microclimatic variability. The detection of permafrost in these areas can be based on *direct* methods such as digging, core drilling, temperature measurements in boreholes, or mapping of landforms containing ground or glacier (e.g. rock glaciers, ice-cored moraines). Drilling in mountain permafrost areas is difficult and expensive but provides direct and detailed information at the drill site (cf. LACHENBRUCH et al. 1988, VONDER MÜHLL & HOLUB 1992, ISAKSEN et al. 2001). Some indirect methods of detecting permafrost include mapping of BTS (Base temperature of winter snow cover, cf. HAEBERLI 1973, HOELZLE 1996) and interpretation of geophysical measurements.

A number of physical parameters, such as seismic velocity, electrical conductivity etc., change during phase change of water from unfrozen to frozen state. This change of physical properties offers possibility of identifying frozen ground and determining its characteristics by using geophysical sounding methods (e.g. VONDER MÜHLL et al. 2002). DC resistivity soundings have shown to be a reliable tool to detect permafrost and ground-ice characteristics. The results give indications of permafrost thickness and different types of ice (e.g. BARSCH 1973, FISCH et al. 1977, HAUCK 2001). Since the 1960's resistivity soundings have been applied in the arctic (e.g. OSTERKAMP et al. 1980) and in high-mountain areas (e.g. KING 1982). Much of the work in high mountain areas is related to rock glaciers (e.g. BARSCH 1973, EVIN & FABRE 1990, VONDER MÜHLL 1993) and ice-cored moraines (e.g. ØSTREM 1964, FISCH et al. 1977, HAEBERLI & EPIFANI 1986, EVIN & FABRE 1990, KNEISEL et al. 2000), using mainly one-dimensional depth soundings. Later, soundings on permafrost sites without morphological expressions of permafrost were carried out in Norway (e.g. ØDEGÅRD et al. 1996). DC-resistivity data can also be processed with algorithms that allow a two-dimensional tomography section of the subsurface (e.g. LANZ et al. 1997, VONDER MÜHLL et al. 2002). The latter has proved to be of high interest in mapping and understanding of the transition area of alpine permafrost (e.g. HAUCK 2001, VONDER MÜHLL et al. 2002). Multi-temporal resistivity soundings together with temperature measurements give an indication of ground-ice build-up and melting during the year (e.g. HAUCK 2001). Today, DC-resistivity measurements are an important tool in investigating glacial and periglacial landforms (e.g. BERTHLING et al. 1998, ISAKSEN et al. 2000, KNEISEL et al. 2000) and in validating numerical or empirical permafrost distribution models (e.g. ETZELMÜLLER et al. 2001, HOELZLE et al. 2001).

This paper summarises 1D DC-resistivity sounding measurements in or near the permafrost zone in southern Norway (Fig. 1). Most of the soundings were taken in high-mountain areas lacking morphological expressions of permafrost. These measurements were carried out as part of a regional permafrost-mapping project, which resulted in an empirical permafrost distribution model based on topographic and meteorological information (ETZELMÜLLER et al. 1998, ETZELMÜLLER et al. 2003). The objective of the paper is to assess the feasibility of the method for permafrost mapping and model validation in areas with lack of morphological permafrost indicators, and show the data interpretation from various sites in southern Norway.

2   Setting

The study area comprises the high-mountain areas of southern Norway (Fig. 1). The Scandinavian mountain chain stretches SW-NE parallel to the Atlantic coast and consists of pre Cambrian basement partly covered by the Caledonian nappes. The tectonic South Norway

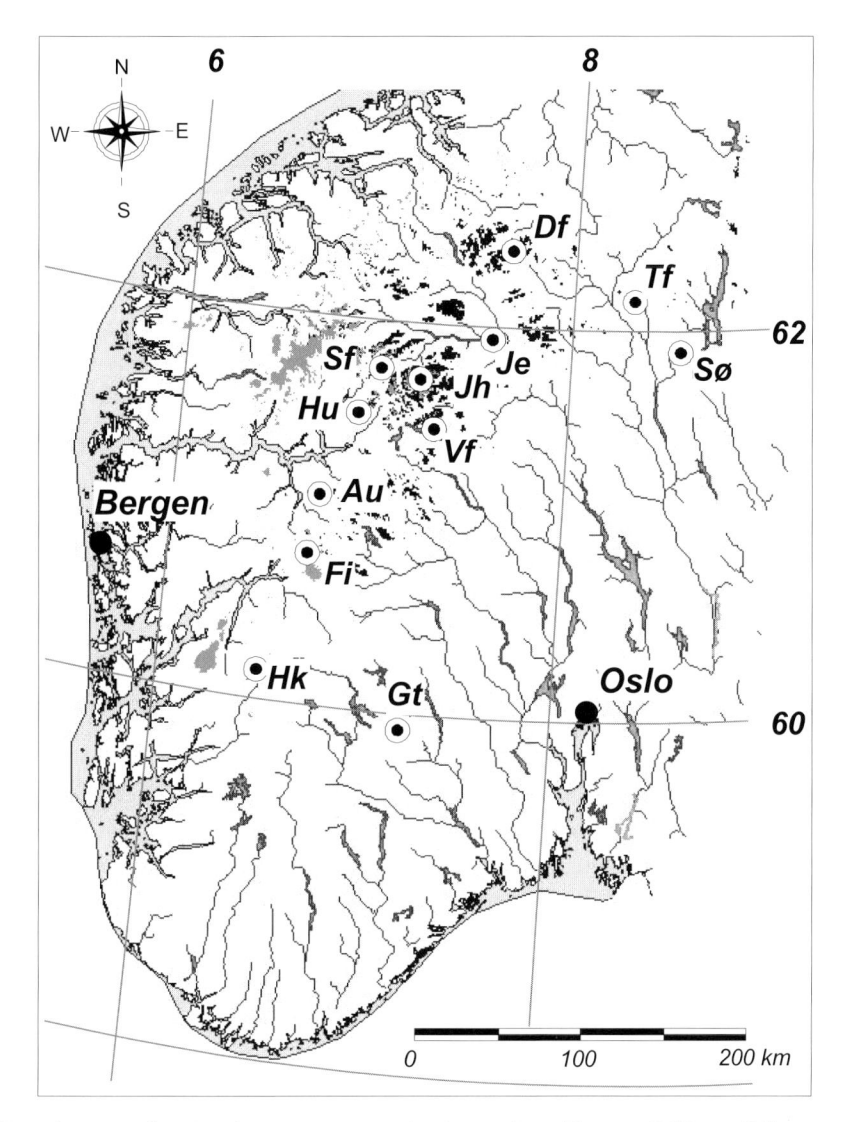

Fig. 1. Location map showing the measurement sites in southern Norway. Df-Dovrefjell mountain, Tf-Tron mountain, Sø-Sølen massif, Je-Jetta mountain, Sf-Sognefjell mountains, Je-Jetta mountain, Jh-Juvvass/Jotunheimen, Hu-Hurungane, Vf-Valdresflya, Au-Aurland mountains, Fi-Finse area, Hk-Haukelifjell mountains, Gt-Gaustatoppen mountain. Grey shade are glaciers, black shading are probable permafrost areas according to ETZELMÜLLER et al. (2003).

(between 58N05E and 63N12E) is dominated by the South Scandinavian dome (LIDMAR-BERGSTRØM 1999), an area uplifted during Tertiary time. The highest mountain areas are close to the western coast, decreasing in altitude towards the east. The area has been subjected to peneplanation since the Mesozoic, linear erosion during the Tertiary and relief enhancement during the Pleistocene glaciations. This led to widespread relatively low-relief mountain areas, intersected by deep incised glacial valleys.

According to an empirical regional model (ETZELMÜLLER et al. 1998) mountain permafrost in southern Norway is mainly concentrated in a 50 to 100 km wide zone between Hallingskarvet in the south and the Dovrefjell mountains in the north (Fig. 1), covering app. 3000 km^2, which is nearby double the present glacier area. Only small areas east and west of this zone have permafrost. On the western side, glaciers normally cover high mountain areas. Furthermore, the lower permafrost limit rises to over 1600 m a.s.l. due to an increasing maritime influence. Thus, there are few areas where permafrost can exist. On the eastern side of this zone only small mountain areas or single peaks reach altitudes above 1400 m a.s.l. In the eastern areas these peaks seems to have permafrost. Preliminary results of a permafrost distribution modelling applied to the whole of Scandinavia and new field investigations indicate even lower alpine permafrost limits, down to below 1300 m a.s.l. in the Femund-Sølen area (Fig. 1, HEGGEM et al. 2003). Much of the permafrost is situated in areas covered by coarse block fields or blocky ground moraines (cf. SOLLID & SØRBEL 1979). A pre-glacial age of these deposits has been suggested, based on vegetation successions (DAHL 1956), warm-climate clay weathering remnants (e.g. ROALDSET et al. 1982) and a regional distribution above glacial trimlines (SOLLID & SØRBEL 1979, NESJE et al. 1988) or in areas of possible landscape preservation beneath cold-based part of the inland ice sheet (SOLLID & SØRBEL 1994).

Basic information about the study sites are summarised in Table 1. All sites are situated close to or above the regional lower permafrost limit, and most of them are situated in block fields or other block-rich surfaces.

3 Methods

For theory and detailed introduction of the resistivity sounding method we refer to standard textbooks (e.g. REYNOLDS 1997). The principle of the method is based on different types of material having different ability to conduct electricity. The resistivity ρ (in Ωm) is the reciprocal of electrical conductivity and defined as the electrical resistance of a cylinder with unit cross-section area and length. The resistivity of geological material depends on the conductivity of mineral grains, the chemical composition and amount of water filling the pore spaces, and temperature. The application of resistivity soundings in permafrost mapping studies is based on the large contrast of water resistivity ($\sim 10-100$ Ωm) in relation to ground ice (10^3-10^6 Ωm) or glacial sedimentary ice ($10^6- >10^7$ Ωm). This means that the method has the potential to detect ground ice layers and thus permafrost in loose, water-rich material. On dry bedrock-dominated sites, however, the method has clear restrictions as especially dense, ingenious and metamorphic rock types have comparable resistivities to ground ice.

In field we measure the resistivity of the ground by transmitting a controlled current [I] between the current electrodes, and measure the potential [U] between two other electrodes. The resistance [R] is calculated using Ohm's law: R = U/I. The apparent resistivity (ρ) is calculated following $\rho = K(U/I)$, with K being the geometrical factor. In this paper only the symmetrical Schlumberger-configurations were shown and discussed. For the Schlumberger configuration, which is used in this study (see Fig. 2), K is defined as (REYNOLDS 1997)

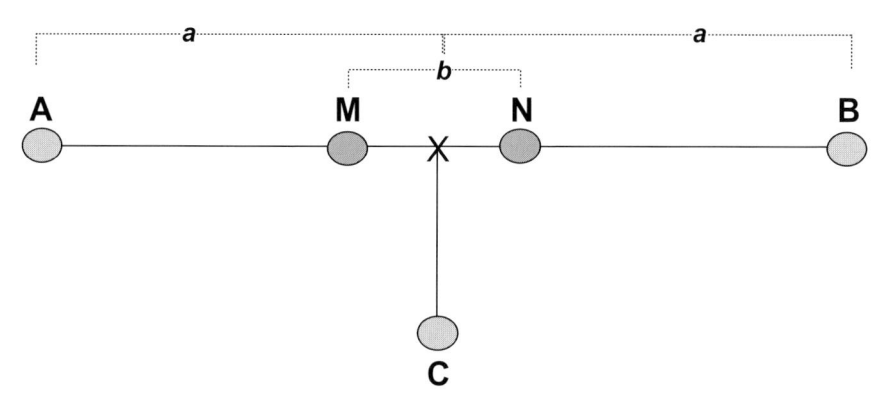

Fig. 2. Measurement array. In this study Schlumberger arrays were measured. Each profile of 400 m length (AB/2) contains 25 data points including 6 double measurements when changing potential electrode spacing. The distances of the current electrodes increased logarithmically during the measurements.

$$K = \frac{2\pi a^2}{b}\left[1 - \frac{b^2}{4a^2}\right]; a \geq 5b.$$ Here, a is the distance between the potential electrodes and the sounding centre, and b is the distance between the potential electrodes. Fig. 2 shows the set-up of the soundings used in this study.

In our study, an ABEM Terrameter was used, measuring Schlumberger configuration with maximum AB/2-distances usually on the range of 200 m to 400 m (see Fig. 2). Steel and aluminium rods were used as potential and current electrodes, respectively. On every main site at least two profiles were recorded, in different altitude levels. All sites were flat or the AB direction was chosen to follow, more or less, a contour line. Also snow patches, which interfere heavily with measurements and stability, were avoided. In total, 35 soundings at 12 different sites (Fig. 1) are presented here. All sites were located in high-mountain areas dominated by block fields, course moraines and regolite, thus high amounts of unfrozen water in possible permafrost and saline pore water were avoided.

From the measurements, field curves were established by plotting ρ against the distance between the current electrode and the profile centre (AB/2) on a log-log scale (see also Figs. 5–8). These curves can be interpreted either qualitatively or quantitatively (see also REYNOLDS 1997). Qualitative methods include the interpretation of the curve shape (Fig. 3). If we assume a situation with three sub-surface layers, one can generate four different curve shapes (Fig. 3, Types "H", "A", "K", "Q" REYNOLDS 1997). E.g., ice-rich permafrost sites normally generate a bell-shaped curve of type "K", where the low-resistivity active layer overlays the high-resistivity permafrost layer, followed again by a low-resistivity sub-permafrost layer. A shape analysis thus gives a first impression about resistivity distribution with depth. Quantitative methods include computer modelling of the data. Here, we can generate either automatically or manually a synthetic curve describing the resistivity variation with depth. The algorithm then tries to calculate an optimal curve giving a best possible fit of the synthetic curve to the field data in an iterative process (inversion technique). In this study

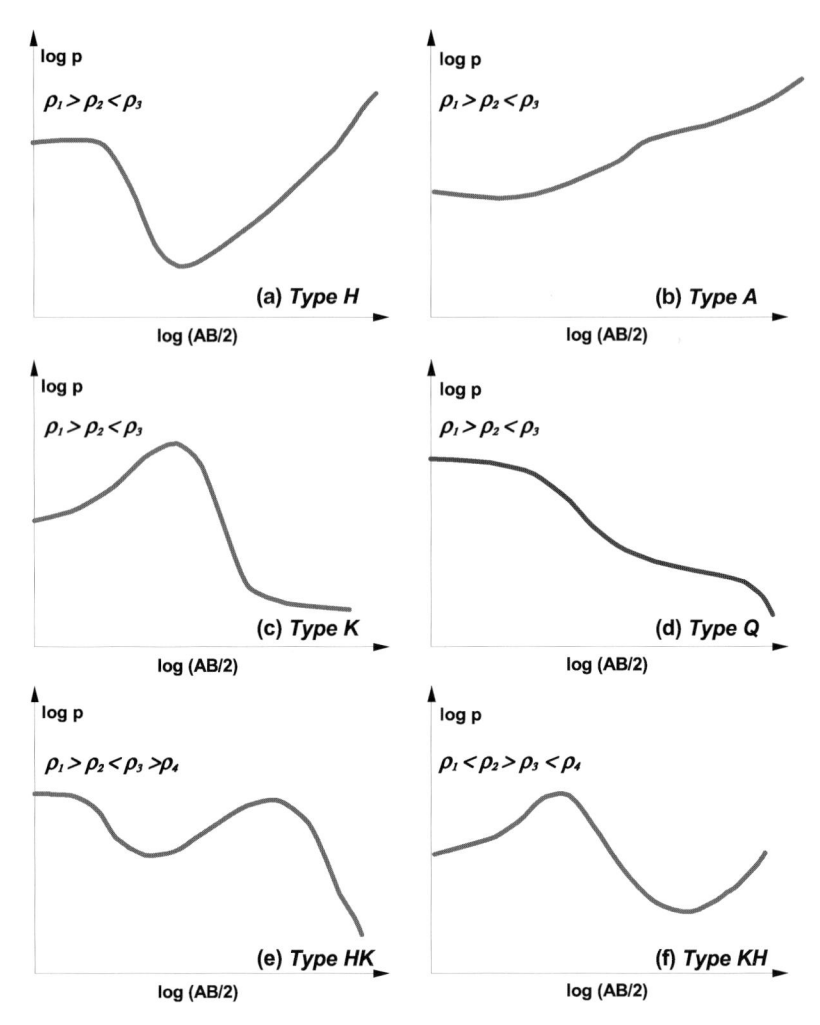

Fig. 3. Typical apparent resistivity curve shapes, observed for three-layer (a–d) or 4-layer cases (e–f). ρ_1 is the resistivity for the first layer. The figure is adapted from REYNOLDS (1997), slightly changed.

the profiles were interpreted using the software RESIXPlus (Interpex Lim). Initially, for all sites at least a 3-layer model was assumed, and for some sites a 4-layer model was used. A quantitative analysis includes the principle of equivalence. This means, that relatively thicker layer of lower resistivity or thinner layers of higher resistivities can produce the same apparent resistivity values. For all profiles an analysis of equivalence was performed, and the result displayed as a separate diagram in Figs. 5 to 8. Noisy or erroneous field data will produce high variation of the relation between apparent resistivity and layer thickness. No constraints were manually applied to the inversions. Large differences between segments due to the change of potential electrode distance were shifted manually within the software. Outliers were masked before the inversion in a few models.

4 Results

As a generalisation, four major surface types were distinguished based on the field conditions (Fig. 4):
- Type I: Block field dominated by boulders with diameter of more than 50 cm and up to several metres
- Type II: Block and /or gravel-rich ground moraine and block fields consisting of finer-grained blocks with diameter below 50 cm
- Type III: Extensive exposed bedrock with a thin cover of unconsolidated material.
- Type IV: Ground moraine, sandy to silty matrix, covered by vegetation (above tree limit)

Permafrost sites were indicated by the measurements in the first two types of surfaces (Figs. 5 and 6). Reliable identification of permafrost was not possible in the third (Fig. 7). The limitation of resistivity soundings on bedrock for permafrost identification is well known, and more thoroughly discussed in e.g. Hauck (2001). Calibrating profiles in probably no-permafrost terrain comprises the Type IV-sites (Fig. 8). These measurement displays a sort of "background" resistivities for the particular sites, meaning unfrozen bedrock or sediment resistivities. Probable bedrock resistivities at the given sites are also shown in Table 1.

4.1 Coarse block fields

The background resistivities in most of the sites were measured to be around 10–30 kΩm. In coarse-grained block fields bell-shaped curves (Type K) with high resistivities (50–>100 kΩm) indicate permafrost of below 50 m thickness (Figs. 5a to c). Blocks close to the surface have air-filled voids, and often produce high resistivities in the active layer, and a weak increase of resistivity with depth is measured (e.g. Fig. 5c). An initial decrease with a following bell-shaped curve indicates a low-resistivity layer above the permafrost layer (Fig. 5b, Type HK, see Reynolds 1997). These layers can be interpreted as a water-rich horizon, possibly originating from thick active layers (3–5 m) or sub-surface runoff, which cannot penetrate into the permafrost. Such a type of sounding is shown in Fig. 5b from the Hurungane mountain. Block-fields on non-permafrost ground are characterized by a steep decrease of resistivity with depth below the block limit (< 10 kΩm, Fig. 5d), producing a Type H curve. This site in the Aurland mountains is very snow-rich, witnessed by numerous snow fields, and thus ground thermal regime is heavily influenced by a thick snow cover. Steep rises of curves indicate extreme high resistivity just below the surface, resulting in noisy data and measurement errors. Fig. 5e displays sounding data over an ice-cored moraine on the Dovrefjell mountains, suggesting sedimentary glacier ice (1–2 MΩm) about 10 m thick. High resistivities of over 200 kΩm are measured at high altitudes of Gaustatoppen, indicating thick ice-rich permafrost (Fig. 5f). High ice content filling numerous joints in the local bedrock and large permafrost thicknesses (> 100 m) is verified at this site during the construction of a tunnel in the mountain.

Except for the site over the ice-cored moraine (Fig. 5e) RMS errors were below 8.5%, resulting in a reasonable fit of the curve with the data points. Also the equivalence analysis displays small variations around the best-fit model.

Table 1. Description of measurement sites and permafrost limits from other studies. The localities are indicated in Fig. 1.

Place	Alt. of sound. (m a.s.l.)	Bedrock (estimated resistivity range)	Site description and surface cover type	Related permafrost studies and estimated permafrost limits
Finse – Jomfrunut/ Sandalsnut/ Finsenut	1300–1600	Phyllite $\sim <1$–$10\ \text{k}\Omega\text{m}$	Low-relief area (paleic surface), glacially scoured. Thin cover of weathered phyllite, silty to sandy matrix. Type III sites. Snow-rich area.	No detailed permafrost studies in the area, expected limit at >1600 m a.s.l. (Etzelmüller et al. 1998). Snow patch-related permafrost abundant below this limit.
Tronfjell/ Flattron	1100–1650	Gabbro ~ 4–$9\ \text{k}\Omega\text{m}$	Mountain massive surrounded by deep valleys. Coarse block fields above 1350 m a.s.l. (Type I), ground moraine below 1300 m a.s.l. Dry area, exposed to snow-drift.	BTS-measurements (Engelien 1995) indicate permafrost in the summit area. Probably strong inversion effects, giving lower temperatures towards the valley.
Jetta/Blåhøi mountain	1550	Quartzite, shist $\sim 10\ \text{k}\Omega\text{m}$	Mountain massive surrounded by deep valleys. Partly coarse block field, sorting (Type II). Area is exposed to snow-drift.	BTS-measurements indicate permafrost conditions at least above 1500 m a.s.l.(Bø 1998). There are probably strong inversion effects.
Gausta mountain	1100–1600	Quartzite, ~ 15–$70\ \text{k}\Omega\text{m}$	Mountain massive surrounded by deep valleys. MAAT on summit meteorological station was $-4°\text{C}$. Block field (Type I) above 1400 m, Ground moraine (Type IV) below 1300 m a.s.l. The area is exposed to snow-drift	Permafrost discovered above 1600 m within the mountain during construction of a tunnel for military installations. Resistivity soundings indicate permafrost down to 1500 m a.s.l.
Haukeli mountains	1450–1550	Metamorphic bedrock (e.g. gneiss) $\sim 15\ \text{k}\Omega\text{m}$	Low-relief area, paleic surface, glacially scoured, low material cover (Type III). Snow-rich area.	No detailed permafrost studies in the area, expected limit at >1600 m a.s.l. (Etzelmüller et al. 1998). Soundings do not indicate permafrost.
Valdresflya	1520	Sparagmite 15–25 $\text{k}\Omega\text{m}$	Flat mountain plateaus (paleic surface). Block-rich ground moraine (Type II). Snow-rich area.	No detailed permafrost studies in the area, expected limit at <1500 m a.s.l. (Etzelmüller et al. 1998). Resistivity soundings do indicate possible permafrost.
Sognefjell mountains	1520–1620	Shists, phyllite ~ 8–$20\ \text{k}\Omega\text{m}$	Low-relief area, paleic surface, glacially scoured, low material cover (Type III). Snow-rich area, extensive snow patches during summer	No detailed permafrost studies in the area, expected limit at >1600 m a.s.l. (Etzelmüller et al. 1998).

Table 1. (cont.)

Place	Alt. of sound. (m a.s.l.)	Bedrock (estimated resistivity range)	Site description and surface cover type	Related permafrost studies and estimated permafrost limits
Hurrungane mountains	1600–1700	Gabbro/ultramafic rocktypes ~15 kΩm	High alpine mountain massive west of Jotunheimen, cirque glaciation. Block fields above 1500 m a.s.l. (Type I). Snow-rich area.	No detailed permafrost studies in the area, expected limit at > 1600 m a.s.l. (ETZELMÜLLER et al. 1998). Resistivity soundings do indicate permafrost.
Aurland mountains	1500–1620	Shists 8–10 kΩm	Low-relief area, paleic surface. Block field (Type I and II) above bedrock. Snow-rich area, extensive snow patches.	No detailed permafrost studies in the area, expected limit at > 1650 m a.s.l. (ETZELMÜLLER et al. 1998). Resistivity soundings did not indicate permafrost.
Dovre mountains	1400–1700	Sandstone/Quartzite < 10 kΩm	Low-relief area with scattered alpine mountains. Mainly block fields and thick ground moraine cover (Type II). Extensive patterned ground.	Well-investigated area with shallow bore-holes, BTS and DC resistivity soundings (ØDEGÅRD et al. 1996, SÆTRE 1997, ISAKSEN et al. 2002, SOLLID et al. 2003). Lower permafrost limits at 1450 m, locally lower on wind-exposed sites.
Juvvassflya/Jotunheimen	1400–1800	Gabbro < 10 kΩm	High plateaux surrounded by alpine mountain topography. Block fields and thick ground moraine cover (Type II). Extensive patterned ground.	Well-investigated area with deep bore hole (SOLLID et al. 2000, ISAKSEN et al. 2001), BTS, DC resistivity sounding (ØDEGÅRD et al. 1996, ISAKSEN et al. 2002) and hammer seismics (HAUCK et al. 2001). Lower permafrost limit at 1450 m a.s.l.
Sølen mountain	1300–1050	Quartzite, Conglomerate > 50 kΩm	Mountain massif surrounded by mountain plateaux. Bouldery block field down to 1100 m a.s.l. (Type I). Coarse, sandy ground moraine below this limit (Type IV). Dry area, exposed for snow drift.	Investigated area with BTS-measurements, Hammer seismics and DC resistivity soundings (e.g. HEGGEM et al. 2003, JULIUSSEN 2003). Strong aspect-dependency of permafrost distribution, with limits of ca. 1050 m a.s.l. on northerly slopes.

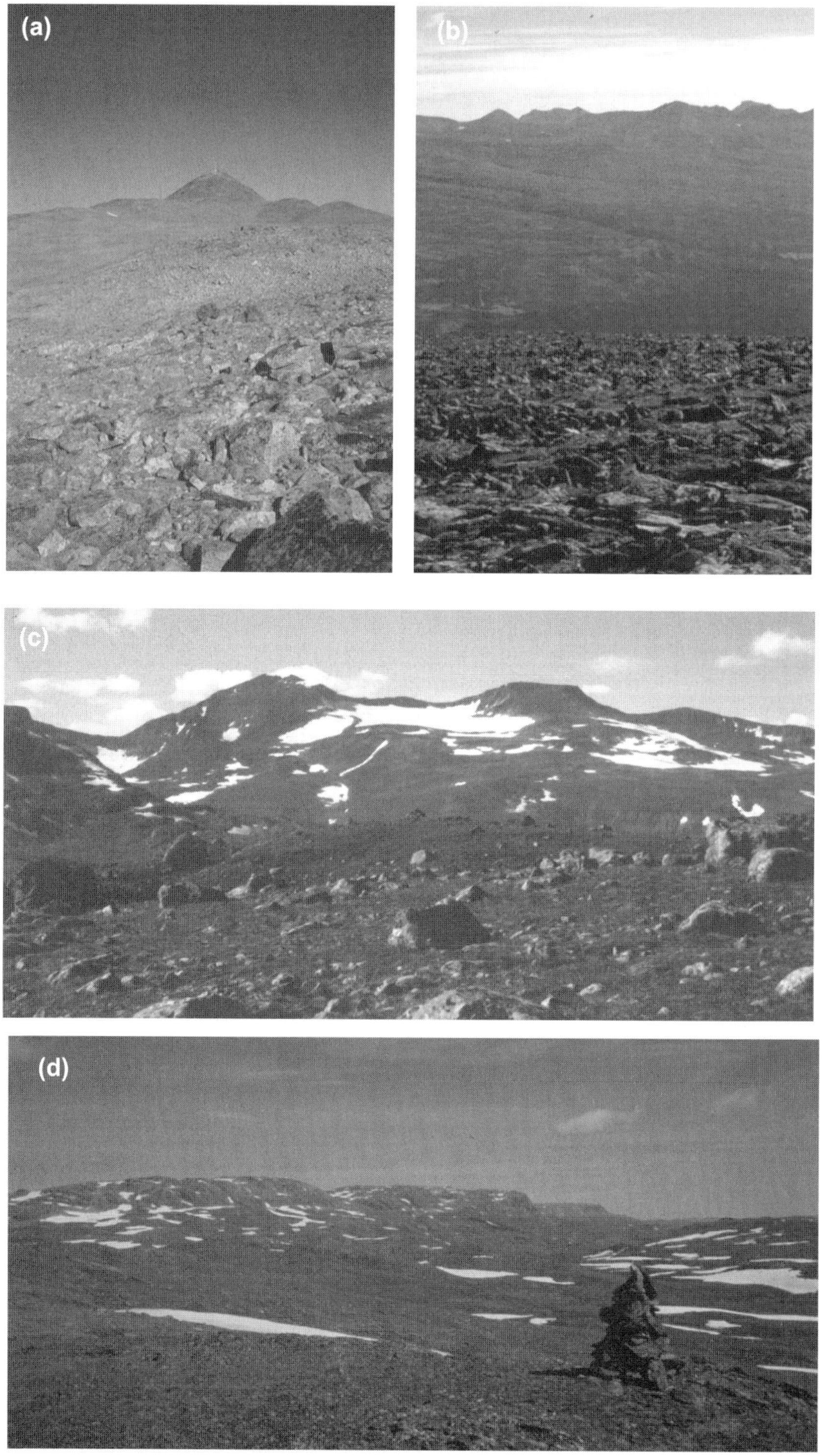
(a)
(b)
(c)
(d)

4.2 Block-rich ground moraine and finer block fields

Block-rich ground moraines and finer-grained block fields show generally lower resistivities in the upper-most layers ($<\sim 20$ kΩm), because of a smaller amount of air-filled voids and possible unfrozen pore water. The sounding from Jotunheimen (Fig. 6a) was made just above the lower limit of permafrost (e.g. ISAKSEN et al. 2002), and shows a low-resistivity layer above a higher one (~ 100 kΩm), describing a Type *H* curve. This is again interpreted as thick water-rich active layer above permafrost. The site on Dovrefjell (Fig. 6b) is situated close to the lower permafrost limit according to SÆTRE (1997), ISAKSEN et al. (2002) and SOLLID et al. (2003), dominated by morainic cover sediments and frost sorting. The sounding shows a 10 m thick high resistivity ice-rich permafrost (~ 100 kΩm) layer, overlaying probably partly frozen bedrock. The site on Valdresflya (Fig. 6c) on blocky, dry ground moraine (Fig. 4c) shows high upper-layer resistivities (>10 kΩm), then a small increase before falling at greater depths. The interpretation suggests a dry active layer, overlaying shallow permafrost with app.10 m thickness. However, the curve could also be produced by dry ground-moraines overlying bedrock. Valdresflya is a snow-rich and wet area, so the first interpretation seems more probable. Also the permafrost map by ETZELMÜLLER et al. (1998) indicates permafrost at that altitude in the area. On the Jetta mountain (Fig. 4c), the sounding was made on a north-facing slope setting, indicating shallow (<10 m), relatively high-resistivity permafrost (~ 100 kΩm) (Fig. 6d, Type *KH*). All soundings in this type of surface show a reasonable RMS error. The equivalence analyses display a good concentration around the best-fit model, especially for the deeper layers.

4.3 Thin material cover and exposed bedrock

Resistivity measurements on sites with thin surface sediment cover and dense bedrock also produce relatively low resistivities in relation to coarse weathering material. Sounding data commonly show constant or slightly rising resistivities with depth (Fig. 7a, 7c) of type *A*, and do not give clear indications about the presence of permafrost. Fig. 7b shows a sounding made on a mountain top north of the Finse valley. The terrain is similar to that shown in Fig. 4d. The sounding here shows a clear bell-shaped form (Type *K*), with low resistivities (10 – 12 kΩm) in the uppermost layers. The site is wind-exposed with low snow cover, thus permafrost is possible. In addition, the bedrock is dominated by jointed phyllites, thus ice lenses are possible, producing the bell-shaped form of the resistivity curve. However, resistivities are low in comparison to other permafrost areas (~ 10 kΩm), therefore clear conclusions cannot to be drawn here. Fig. 7c display a typical homogenous bedrock profile with low sediment cover, producing a nearby horizontal resistivity curve. Due to lacking or sparse sediment cover, the field data also seems much more noisy, with somewhat higher RMS errors

Fig. 4. Photographs describing the typical surface conditions on the measurement sites (see also Fig. 1). (a) Coarse block-rich surface, Gaustatoppen. (b) finer-grained block fields on Jetta mountain, (c) finer-grained surface, ground moraines and block fields, Valdresflya, (d) sites with low material cover and exposed bedrock, Sandalsnut/Finse area.

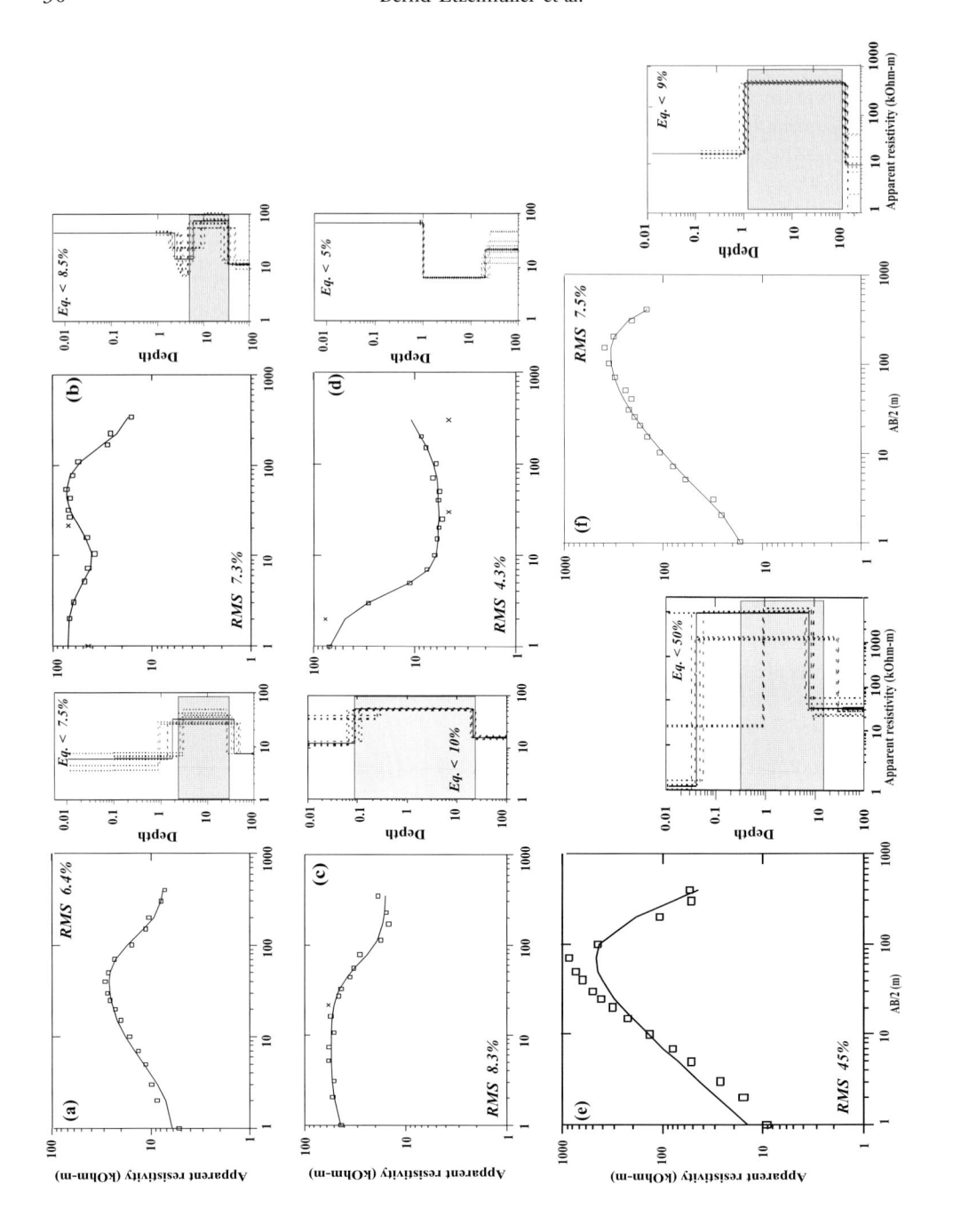

than presented before. This obviously also influence the equivalence analyses, resulting to wider range of resistivity-depth models.

4.4 Vegetation-covered ground moraine above tree limit

Using 1D-resistivity soundings for permafrost detection, profiles in nearby probable non-permafrost terrain should be carried out for calibration purposes. Fig. 8 displays such profiles, mostly taken on levelled ground with dense vegetation (low to mid-alpine) and varying degree of ground-moraine cover. The moraine cover in mountain areas of southern Norway is mostly coarse-grained, resulting in high resistivities in the top-layers (~ 10 kΩm) when measuring during dry summer conditions (Figs. 8a, c). On these sites type *H* or flat type *HK* curves (Fig. 3) were produced, indicating near homogenous or wetter conditions with depth. Fig. 8b displays a situation where the moraine cover seems to have higher resistivities than the top layer and the possible underlying bedrock. This also produces a bell-shaped curve (Type K), which in this case could not be attributed to a permafrost situation. Firstly, the site was below 1100 m a.s.l. and exposed towards south, and thus certainly below the regional permafrost limit in this area. Secondly, the sounding was carried out on a ridge-like feature, introducing geometrical distortion and thus noisy data

5 Discussion

The field soundings and their interpretation contain several errors. As no topographic correction was introduced, the model assumes no topography and horizontal layering. In mountainous terrain the field survey always encounters varying topography with horizontal stretches over 800 m, introducing geometrical distortions. Block field and bouldary moraines often produce difficulties with the connection between electrodes and ground, decreasing the signal to noise ratio of the measurements, and often forcing the operator to measure with too low current. This results in noisy measurements, especially in high contact resistance ground typical for high sedimentary ice content. This is exemplified shown in Fig. 5e. Field resistivity rises much steeper than the maximum angle of slope that a log-log curve may possible have (see also REYNOLDS 1997). This results in large RMS-errors and variation

Fig. 5. Resistivity curves and models obtained from Type I surfaces. The left diagram shows the points measured with different electrode distances ("AB/2") plotted against apparent resistivity (ρ). The solid line is a synthetic curve obtained after the inversion procedure, which gave best fit to the data points (square symbols). Points marked as a "X" were defined as outliers and not included in the inversion procedure. *RMS* ("root means square") denotes the fit error of the synthetic curve to the points, expressed in percentage. The right diagram shows the results of the equivalence analyses by plotting ρ against depth (in m). The solid line denotes the layer model for the highest fit ($= RMS$ error), while the various dashed lines indicate layer models with *RMS* errors between the best fit-line and a given percentage, which is given as ("*Eq.*") in the graphs. The possible permafrost layer is shaded. (a) Gausta mountain, 1500 m a.s.l. (b) Hurrungane mountains, 1700 m a.s.l., (c) Hurrungane mountains, 1600 m a.s.l., (d) Aurland mountains, 1620 m a.s.l., (e) Dovrefjell mountain, ice-cored moraines (~ 1650 m a.s.l., (f) Gausta mountain 1650 m a.s.l.

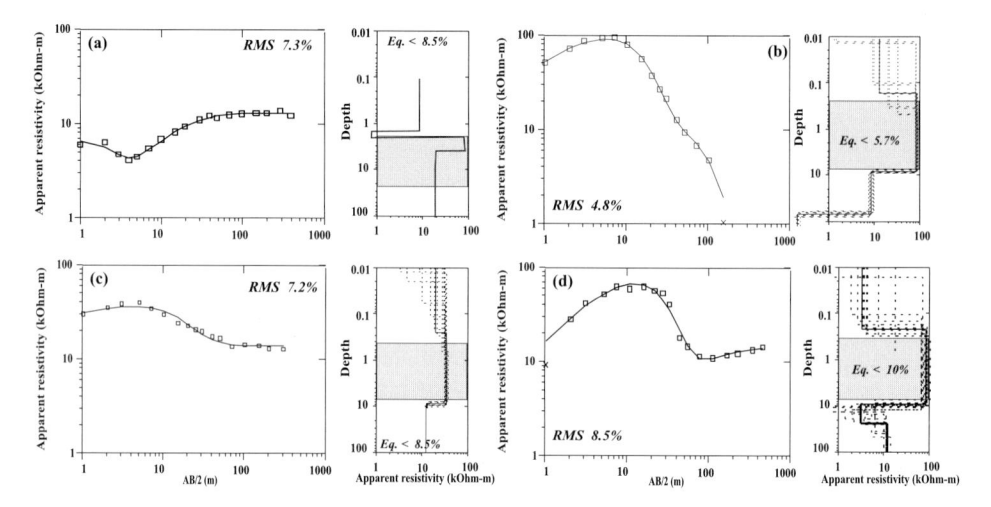

Fig. 6. Resistivity curves and models obtained from Type II surfaces. For explanation, see Fig. 5. (a) Juvvass area/Jotunheimen, 1550 m a.s.l., (b) Dovrefjell mountain, 1450 m a.s.l, (c) Valdresflya, 1520 m a.s.l., (d) Jetta mountain, 1550 m a.s.l. The possible permafrost layer is shaded.

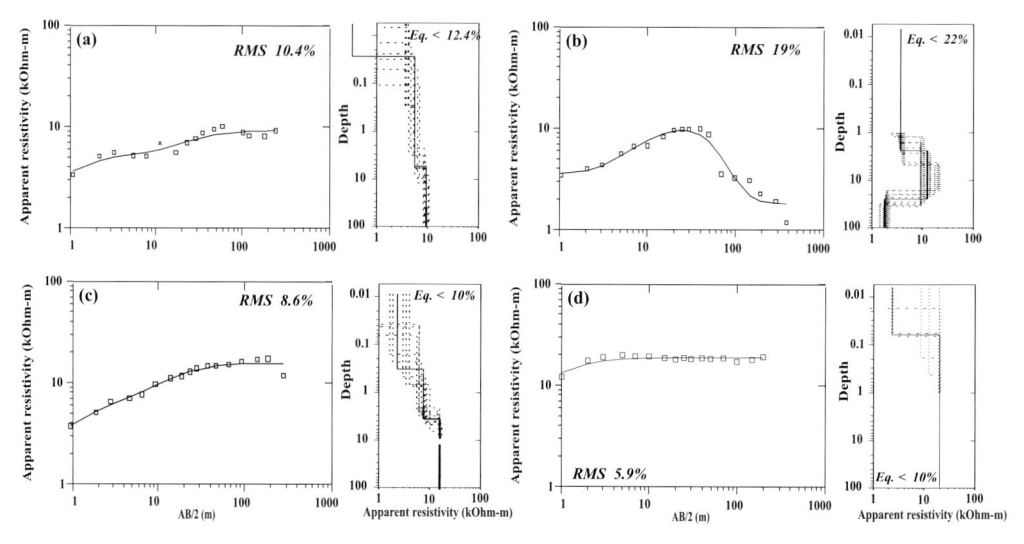

Fig. 7. Resistivity curves and models obtained from Type III surfaces. For explanation, see Fig. 5. (a) Sognefjell mountains, 1620 m a.s.l., (b) Finsenut, 1550 m a.s.l., (c) Haukelifjellet, 1450 m a.s.l. (d) Sognefjell mountain 1520 m a.s.l.

in the possible layer thickness-resistivity relation (equivalence), clearly suggesting errors in field measurements or geometric effects. In our case the profile was measured over a sharp moraine ridge, containing an ice-body. Therefore, absolute resistivity values obtained in such cases may be erroneous with several orders of magnitude, and thus absolute resistivity values must be

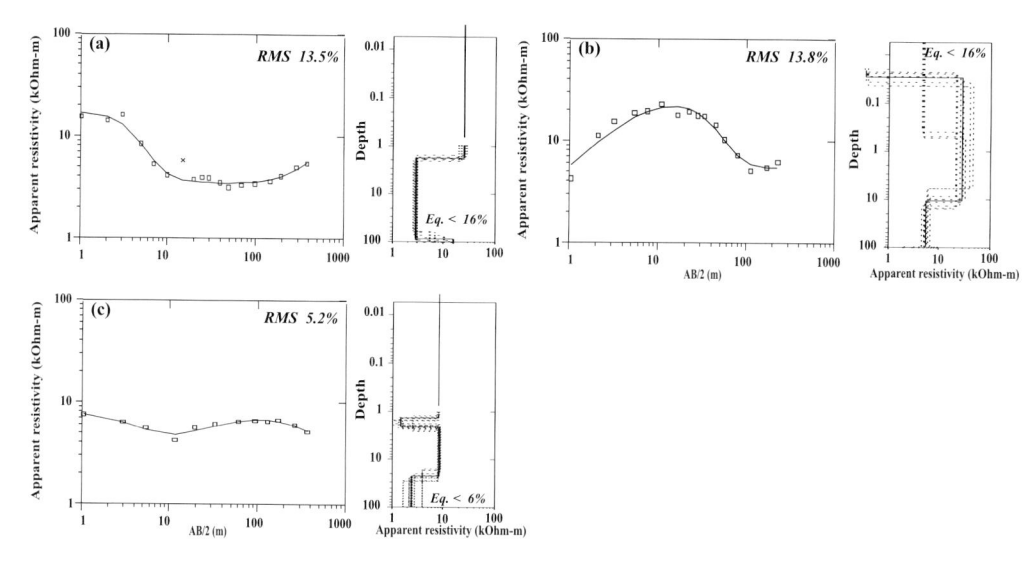

Fig. 8. Resistivity curves and models obtained from Type IV surfaces. For explanation, see Fig. 5. (a) Tron mountain, 1300 m a.s.l., (b) Gausta mountain, 1100 m a.s.l., (c) Finse valley, 1220 m a.s.l.

used with caution. Generally, most of the soundings from this study show reasonable RMS values, of mostly below 10%. Also the equivalence analyses show a good concentration around the best-fit line, indicating relatively consistent measurements. However, the mentioned errors are common for all measurements carried out in high-mountain areas, producing comparable results and sounding patterns. Soundings on rock glaciers from different parts on Svalbard (e.g. BERTHLING et al. 1998, ISAKSEN et al. 2000) and the Alps (e.g. FISCH et al. 1977, EVIN & FABRE 1990, VONDER MÜHLL 1993) show comparable results. With the rising number of this type of DC resistivity investigations in high-alpine environments the comparability of the results from such soundings has been increased considerably.

Generally, 1D-DC resistivity soundings are not a stand-alone tool for determining permafrost. On landforms indicating the presence of permafrost, such like rock glaciers (e.g. BARSCH 1992) or ice-cored moraines/large-scale push moraines (e.g. ETZELMÜLLER et al. 1996), the method may give an indication of varying ice-content and ice characteristics as sedimentary glacier ice often shows much higher resistivities than ground ice (e.g. FISCH et al. 1977, HAEBERLI & VONDER MÜHLL 1996). For areas with no morphological expression of permafrost occurrence, the method must be regarded as a supplement in combination to other techniques, such like seismic surveys or BTS measurements. As shown in this paper, bell-shaped curves of type *K* are also produced on bedrock settings and on probable dry sediments. However, a high number of soundings like those described in this paper seem to reveal an interpretable pattern, especially in the block-rich areas. In many of the sites presented here additional information exists (see Table 1), confirming the presence of permafrost.

6 Conclusions

In this study 1D DC-resistivity sounding allowed us to distinguish different patterns of resistivity curves. Possible permafrost in extensive boulder fields or blocky ground moraine reaches resistivities of around 100 kΩm, which indicates relatively ice-rich permafrost. The curves show a typical bell shaped curve (Type K) when permafrost is present. At several sites, a thick low resistivity layer above the supposed permafrost table was observed, indicating a thick active layer. Boulder fields with and without permafrost were easily distinguished, as the latter produce type H or type Q curves. Sites with exposed bedrock are difficult to interpret, because they possibly lack larger ice bodies at depth. Dry coarse-grained sediment packages such as ground moraines can produce bell-shaped curves as well, making interpretation in relation to permafrost difficult. The method of 1D resistivity soundings must be regarded as a supplement to other geophysical or geomorphological investigations, and is not satisfying as a stand-alone method in permafrost mapping research.

Acknowledgements

The DC resistivity soundings discussed in this paper were carried out during the years 1993, 1995, 1997 and 2001. The measurements were a part of a permafrost mapping effort in southern Norway, originally initiated by J. L. Sollid, Dept. of Physic. Geography, University of Oslo. One of the soundings in Jotunheimen displayed in this paper (Fig. 6a) was conducted during a NORFA-financed research course ("Mapping and distribution modelling of mountain permafrost") headed by J. L. Sollid and B. Etzelmüller in 2001, under supervision of M. Hoelzle, University of Zurich. J. Fjellanger and E. Brox carried out the measurements in the Finse area (Figs. 7b, 8c) during a MSc. field course arranged by the Dept. of Physical Geography, University of Oslo. Help in the field were received by E. Engelien, O. Berthling, G. Lothe, S. Sætre, H. Konnestad, E. Gudevang, E. S. F. Heggem, H. Juliussen and M. Bø. Parts of these data were included in the Cand.scient,.-degrees of M. Bø (Bø 1998), E. Engelien (Engelien 1995) and S. Sætre (Sætre 1997). The paper was considerably improved by the constructive comments by E. Heggem, B. A. Hobbs, R. J. Bisdorf and A. Hördt. The Department of geology, NTNU, Trondheim and the Department of Geology, University of Oslo, contributed with geo-electrical equipment. We want to thank all mentioned persons and institutions.

References

Barsch, D. (1973): Refraktionsseismische Bestimmungen der Obergrenze des gefrorenen Schuttkörpers in verschiedenen Blockgletschern Graubündens. – Z. Gletscherkd. Glazialgeol. **9**: 143–167.

Barsch, D. (1992): Permafrost creep and Rockglaciers. – Permafrost Periglac. Process. **3**: 175–188.

Berthling, I., Etzelmüller, B., Eiken, T. & Sollid, J. L. (1998): The rock glaciers on Prins Karls Forland, Svalbard (I). Internal structure, flow velocity and morphometry. – Permafrost Periglac. Process. **9**: 135–145.

Bø, M. R. (1995): Permafrost-studier på Jetta i Nord-Gudbrandsdalen, Midt Norge. – Cand. scient. Thesis, University of Oslo, Norway, 127 pp.

DAHL, E. (1956): Biogeographic and geologic indications of unglaciated areas in Scandinavia during the glacial ages. – Bull. Geol. Soc. Amer. **66**: 1499–1519.

ENGELIEN, E. (1995): Blokkmark og permafrost på Tron i Nord-Østerdal, Sør-Norge. – Cand. scient. Thesis, University of Oslo, Norway, 147 pp.

ETZELMÜLLER, B., BERTHLING, I. & SOLLID, J. L. (1998): The distribution of permafrost in Southern Norway; a GIS approach. – In: LEWKOWICZ, A. G. & ALLARD, M. (eds.) (1998): Seventh International Conference on Permafrost, Proceedings. – Collection Nordicana: 251–257; Centre d'Etudes Nordiques, Universite Laval.

ETZELMÜLLER, B., BERTHLING, I. & SOLLID, J. L. (2003): Aspects and concepts on the geomorphological significance of Holocene permafrost in Southern Norway. – Geomorphology, **52**: 87–104.

ETZELMÜLLER, B., HAGEN, J. O., VATNE, G., ØDEGÅRD, R. S. & SOLLID, J. L. (1996): Glacier debris accumulation and sediment deformation influenced by permafrost, examples from Svalbard. – Ann. Glaciol. **22**: 53–62.

ETZELMÜLLER, B., HOELZLE, M., HEGGEM, E. S. F., ISAKSEN, K., STOCKER-MITTAZ, C., ØDEGÅRD, R. S., HAEBERLI, W. & SOLLID, J. L. (2001): Mapping and modelling the occurrence and distribution of mountain permafrost. – Norsk Geograf. Tidskr. **55**: 186–194.

EVIN, M. & FABRE, D. (1990): The distribution of permafrost in rock glaciers of the southern Alps (France). – Geomorphology **3**: 57–71.

FISCH, W., FISCH, W. & HAEBERLI, W. (1977): Electrical D. C. resistivity soundings with long profiles on rock glaciers and moraines in the Alps of Switzerland. – Z. Gletscherkde. Glazialgeol. **13**: 239–260.

HAEBERLI, W. (1973): Die Basis Temperatur der winterlichen Schneedecke als möglicher Indikator für die Verbreitung von Permafrost. – Z. Gletscherkde. Glazialgeol. **9**: 221–227.

HAEBERLI, W. & EPIFANI, F. (1986): Mapping the distribution of buried glacier ice – an example from Lago Delle Locce, Monte Rosa, Italian Alps. – Ann. Glaciol. **8**: 78–81.

HAEBERLI, W. & VONDER MÜHLL, D. (1996): On the characteristics and possible origin of ice in rock glacier permafrost. – Z. Geomorph. N. F., Suppl.-Bd. **104**: 43–57.

HAUCK, C. (2001): Geophysical methods for detecting permafrost in high mountains. – Versuchsanstalt für Wasserbau und Glaziologie, ETH, Mitteilungen **171**: 204 pp.

HAUCK, C., GUGLIELMIN, M., ISAKEN, K. & VONDER MÜHLL, D. (2001): Applicability of frequency- and time-domain electromagnetic methods for mountain permafrost studies. – Permafrost Periglac. Process. **12**: 39–52.

HEGGEM, E. S. F., JULIUSSEN, H. & ETZELMÜLLER, B. (2003): Mountain permafrost in the Sølen massif, Central-Eastern Norway. – 8th Internat. Conf. Permafrost, Proc., Zurich, Switzerland, in press.

HOELZLE, M. (1996): Mapping and modelling of mountain permafrost distribution in the Alps. – Norsk geogr. Tidsskr. **50**: 11–16.

HOELZLE, M., MITTAZ, C., ETZELMÜLLER, B. & HAEBERLI, W. (2001): Surface energy fluxes and distribution models relating to permafrost in European Mountain areas: an overview of current developments. – Permafrost Periglac. Process. **12**: 53–68.

ISAKSEN, K., HAUCK, C., GUDEVANG, E., ØDEGÅRD, R. S. & SOLLID, J. L. (2002): Mountain permafrost distribution on Dovrefjell and Jotunheimen, southern Norway, based on BTS and DC resistivity tomography data. – Norsk Geogr. Tidskr. **56**: 122–136.

ISAKSEN, K., HOLMLUND, P., SOLLID, J. L. & HARRIS, C. (2001): Three deep alpine permafrost boreholes in Svalbard and Scandinavia. – Permafrost Periglac. Process. **12**: 13–26.

ISAKSEN, K., ØDEGÅRD, R. S., EIKEN, T. & SOLLID, J. L. (2000): Composition, flow and development of two tungue-shaped rock glaciers in the permafrost of Svalbard. – Permafrost Periglac. Process. **11**: 241–257.

JULIUSSEN, H. (2003): Kartlegging av permafrost og periglasiale former på Sølen og Elgåhogna, Øst-Norge. – Cand. scient. Thesis, University of Oslo, Norway.

KING, L. (1982): Qualitative und quantitative Erfassung von Permafrost in Tarfala (S-Lappland) und Jotunheimen (N) mit Hilfe geoelektrischer Sondierungen. – Z. Geomorph. N. F., Suppl.-Bd. **43**: 139–160.

Kneisel, C., Haeberli, W. & Baumhauer, R. (2000): Comparison of spatial modelling and field evidence of glacier/permafrost relations in an Alpine permafrost environment. – Ann. Glaciol. **31**: 269–274.

Lachenbruch, A., Cladouhous, T. T. & Saltus, R.W. (1988): Permafrost temperatures and the changing climate. – V. Internat. Conf. Permafrost, Proc., Tapir, Trondheim, Norway, 9–17.

Lanz, E. H. M., Ansorge, J. & Green, A. G. (1997): High-resolution travel-time tomography used to delineate a shallow waste disposal site. – Proc. EEGS-ES '97, Aarhus, 33–36.

Lidmar-Bergstrøm, K. (1999): Uplift histories revealed by landforms of the Scandinavian domes. – In: Smith, B. J., Whalley, W. B. & Warke, P. A. (eds.) (1999): Uplift, Erosion and Stability. – Perspectives on Long-term Landscape Development, 1–7, Geological Society.

Nesje, A., Dahl, S. O., Anda, E. & Rye, N. (1988): Blockfield in southern Norway. Significance for the Weichselian ice sheet. – Norsk Geol. Tidsskr. **68**: 149–169.

Ødegård, R. S., Hoelzle, M., Johansen, K. V. & Sollid, J. L. (1996): Permafrost mapping and prospecting in southern Norway. – Norsk geogr. Tidsskr. **50**: 41–54.

Østrem, G. (1964): Ice-cored moraines in Scandinavia. – Geogr. Ann. **46A**: 282–337.

Osterkamp, T. E., Jurick, R. W., Gislason, G. A. & Akasofu, S. I. (1980): Electrical resistivity measurements in permafrost terrain at the Engineer Creek road cut, Fairbanks, Alaska. – Cold Regions Science and Technology **3**: 277–286.

Reynolds, J. M. (1997): An introduction to applied and environmental geophysics. – John Wiley & Sons, Chichester.

Roaldset, E., Pettersen, E., Longva, O. & Mangerud, J. (1982): Remnants of preglacial weathering in western Norway. – Norsk Geol. Tidsskr. **62**: 169–178.

Sollid, J. L., Holmlund, P., Isaksen, K. & Harris, C. (2000): Deep permafrost boreholes in western Svalbard, northern Sweden and southern Norway. – Norsk Geogr. Tidsskr. **54**: 186–191.

Sollid, J. L., Isaksen, K., Eiken, T. & Ødegård, R. S. (2003): The transition zone of mountain permafrost on Dovrefjell, southern Norway. – 8th Internat. Conf. Permafrost, Proc., Zurich, Switzerland, same volume.

Sollid, J. L. & Sørbel, L. (1979): Deglaciation of western central Norway. – Boreas **8**: 233–239.

Sollid, J. L. & Sørbel, L. (1994): Distribution of glacial landforms in southern Norway in relation to the thermal regime of the last continental ice-sheet. – Geograf. Ann. **76**: 25–35.

Sætre, S. (1997): Undersøkelser av permafrost i Snøhettaområdet, Dovrefjell, Sør-Norge. – Rapportserie i naturgeografi, Geografisk institutt, Universitetet i Oslo **9**: 64 pp.

Vonder Mühll, D. (1993): Geophysikalische Untersuchungen im Permafrost des Oberengadins. – PhD. (Dr.) Thesis, ETH, Zürich, 197 pp.

Vonder Mühll, D., Hauck, C. & Gubler, H. (2002): Permafrost mapping – mapping of mountain permafrost using geophysical methods. – Progress Phys. Geogr. **26**: 643–660.

Vonder Mühll, D. & Holub, P. (1992): Borehole logging in Alpine permafrost, Upper Engadin, Swiss Alps. – Permafrost Periglac. Process. **3**: 125–132.

Addresses of the authors: Dr. Bernd Etzelmüller, Department of Physical Geography, University of Oslo, P.O. Box 1042 Blindern, N-0316 Oslo, Norway. Dr. Ivar Berthling, Norwegian Water Resources and Energy Directorate, P.O. Box 5091 Majorstua, N-0301 Oslo, Norway. Dr. Rune S. Ødegård Gjøvik University College, Department of Technology, P.O. Box 191, N-2802 Gjøvik, Norway.

Electrical resistivity tomography as a tool for geomorphological investigations – some case studies

Christof Kneisel, Würzburg

with 7 figures and 1 table

Summary. Two-dimensional electrical resistivity tomography was used for various geomorphological problems. Field examples from measurements on high resistive, intermediate and conductive material are given. The described case studies illustrate that electrical resistivity tomography is a powerful tool to determine the structure of the subsurface at different resolution depending on the employed array geometry and electrode spacing. For a reliable interpretation of the survey results, knowledge of the geomorphological and geological setting is essential. The application of electrical resistivity tomography on alpine permafrost has shown that the distinct resistivity contrasts in the subsurface facilitate the interpretation. Examples from surveys on more homogeneous conductive fine-grained sediments indicate that the interpretation of electrical resistivity tomograms is made difficult in cases of indistinct resistivity changes in the subsurface. In such cases, electrical resistivity tomography should be used in conjunction with other geophysical techniques to avoid ambiguous results, as they can provide complementary information about the subsurface. For detailed geomorphological/sedimentological studies a combination of electrical resistivity tomography with conventional drillings is recommended.

Zusammenfassung. Die zweidimensionale Widerstandstomographie wurde für unterschiedliche geomorphologische Fragestellungen eingesetzt. Beispiele von Messungen auf sehr resistiven, intermediären und konduktiven Materialien werden dargestellt. Die vorgestellten Fallstudien zeigen, dass die elektrische Widerstandstomographie ein leistungsfähiges Werkzeug zur Charakterisierung des oberflächennahen Untergrundes, bei unterschiedlicher Auflösung in Abhängigkeit von der eingesetzten Elektrodenkonfiguration bzw. des verwendeten Elektrodenabstandes, ist. Für eine zuverlässige Interpretation ist die Kenntnis der geomorphologischen und geologischen Verhältnisse unerlässlich. Die Anwendung der elektrischen Widerstandstomographie im alpinen Permafrost hat gezeigt, dass hier aufgrund der deutlichen Widerstandskontraste im Untergrund die Interpretation erleichtert wird. Anwendungsbeispiele auf homogeneren, leitfähigen feinkörnigen Sedimenten zeigen dagegen, dass die Interpretation der Widerstandstomogramme erschwert wird, wenn nur undeutliche Widerstandsänderungen im Untergrund vorkommen. In solchen Fällen, sollte die 2-D Widerstandstomographie in Verbindung mit anderen geophysikalischen Methoden eingesetzt werden um zweideutige Ergebnisse auszuschliessen, da diese komplementäre Informationen über den oberflächennahen Untergrund liefern. Für detaillierte geomorphologisch/sedimentologische Studien wird eine Kombination mit konventionellen Bohrungen empfohlen.

Introduction

Geophysical methods are particularly suitable for geomorphological investigations since the knowledge of structure, layering and composition of the subsurface at different scales are key parameters for geomorphological problems. Two-dimensional electrical resistivity tomography or electrical resistivity imaging being one of those modern geophysical survey methods which have been used for various environmental studies for some years (e.g. GRIFFITHS & BARKER 1993, DAHLIN 1996, DANOWSKI & YARAMANCI 1999, OGILVY et al. 1999, MOORE et al. 2000, CHAMBERS et al. 2002). The increase of application of modern geophysical methods in geoscience is due to the fact that geophysical methods are fast and non-destructive compared to conventional drilling and to the availability of 2-D inversion software. Furthermore, the more effective data acquisition enabled new fields of application such as subsurface resistivity imaging for monitoring of time-dependent resistivity changes due to for instance changing water contents. In environmental sciences and hydrogeology, such time-lapse resistivity surveys have been successfully applied and include the flow of water through the unsaturated soil zone, but also migration of pollutants and monitoring of dam leakages (e.g. JOHANSSON & DAHLIN 1996, BARKER & MOORE 1998, BENDERITTER & SCHOTT 1999). In agriculture and soil science, geophysical methods may play a more prominent role in future for measuring subsurface structures in high resolution in a non-destructive way (cf. LÜCK et al. 2000, KOSZINSKI & WENDROTH 2001, PANISSOD et al. 2001). Since geoelectrical methods are most suitable for investigating the subsurface with distinct contrasts in conductivity and resistivity, respectively, DC resistivity soundings constitute one of the traditional geophysical methods which are standardly applied in permafrost research to confirm and characterise mountain permafrost for many years (cf. KNEISEL & HAUCK, this volume). During recent years advances have been achieved in using the traditional methods but with more powerful, state of the art instruments and modern data processing algorithms (two-dimensional surveys and data processing, cf. HAUCK & VONDER MÜHLL, this volume). It is emerging that these methods will become increasingly important also in other fields of geomorphology.

The motivation of this contribution is to provide an insight into the broad opportunities and potentials of two-dimensional DC resistivity tomography for (applied) geomorphological studies as shown on different case studies. Furthermore, limitations of interpretation are presented and applicability of different array geometries are discussed.

General principle

Resistivity measurements are made by injecting a direct current (I) into the ground via two current electrodes (A and B in Fig. 1). The resulting voltage difference (ΔV) is measured at two potential electrodes (M and N). The overall purpose of resistivity measurements is to determine the subsurface resistivity distribution. From the current (I) and voltage difference values (ΔV) the resistivity is calculated with

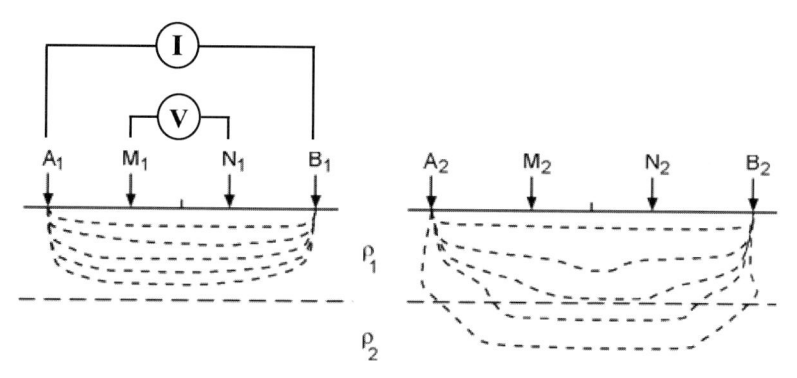

Fig. 1. Conventional four electrode configuration.

$$\rho = K\frac{\Delta V}{I}$$

where K is a geometric factor which depends on the arrangement of the four electrodes. This calculated resistivity value is not the "true" resistivity of the subsurface, but a so called "apparent resistivity". From these measured apparent resistivity values, the resistivity of the subsurface can be derived using different inversion methods implemented in a computer program (see further below).

The basic principle for the successful application of geoelectrical methods in geomorphology/quaternary geology is based on the varying electrical conductivity of minerals, solid bedrock, sediments, air and water and consequently their varying electrical resistivity (cf. Table 1). The resistivity of rock for example depends on water saturation, chemical properties of pore water, structure of pore volume and temperature. The large range of resistivity values for most materials is due to varying water content.

With increasing the spacing between the current electrodes, a deeper penetration depth is obtained (as indicated in Fig. 1) and thus more information about the deeper sections of the subsurface. Resistivity surveys give an image of the subsurface resistivity distribution. Knowing the resistivities of different material types (cf. Table 1), it is possible to convert the resistivity image into an image of the subsurface consisting of different materials. However, as a consequence of overlapping resistivity values of different materials the information might be non-unique.

For electrical resistivity surveys different array types are used. In traditional one-dimensional DC resistivity soundings for geomorphological studies often the symmetrical Schlumberger array is applied, especially if lateral inhomogenities occur. For the interpretation of one-dimensional data the assumption is made that the subsurface consists of horizontal layers and that the resistivity changes only with depth but not horizontally. For the obtained resistivity values a one-dimensional model of the subsurface is interpreted using standard software packages (see KNEISEL 2002 for some examples).

In addition to the Schlumberger array for vertical electrical soundings, another classical survey technique, the resistivity profiling method is used for obtaining lateral changes in

Table 1. Range of resistivities for different materials (compiled mainly after TELFORD et al. 1990).

Material	Range of Resistivity [Ωm]
Clay	$1-100$
Sand	$100 - 5 \times 10^3$
Gravel	$100 - 4 \times 10^2$
Granite	$5 \times 10^3 - 10^6$
Basalt	$10^3 - 10^6$
Schist	$100 - 10^4$
Limestone	$100 - 10^4$
Ground Water	$10 - 300$
Sea Water	0.2
Frozen Ground* / Ground ice* / Mountain Permafrost*	$5 \times 10^3 - 10^6$
Glacier ice (temperate)	$10^6 - 10^8$
Air	Infinity

* Different sources, cf. KNEISEL (1999)

the subsurface resistivity. In this case, the Wenner array is applied where the spacing between the electrodes remains fixed, but all four electrodes are moved for each reading. This gives information about lateral changes, but not about vertical changes in the resistivity.

From the description above the limitations of the resistivity sounding method are evident. The assumed horizontal layering as well as the assumption that the resistivity changes only with depth but not horizontally will not always be found in practice. On heterogeneous ground conditions, the interpretation of one-dimensional soundings can be difficult, as lateral variations along the survey line can influence the results significantly. The sounding curve produces an average resistivity model of the survey area. For some geomorphological studies this might not be problematic and the obtained results from one-dimensional soundings are sufficient. However, individual anomalies will not show explicitly in the results. Two-dimensional resistivity tomography overcomes this problem using multi-electrode systems and two-dimensional data inversion yielding to a more accurate model of the subsurface.

Data acquisition and data processing

Compared to one-dimensional resistivity soundings which typically involve 10 to 20 readings, two-dimensional imaging surveys consist of 100 to several hundred measurements. These data sets are efficiently collected using multi-electrode array systems. The measured apparent resistivities may be used to build up a vertical contoured section showing the lateral and vertical variation of resistivity over the section. The conventional method of plotting the results for the interpretation is the so called pseudosection, which gives an approximate image of the subsurface resistivity distribution. The shape of the contours depend on the array geometry and the subsurface resistivity. The arrays most commonly used for 2-D resistivity surveys are the so called Wenner, Wenner-Schlumberger, Pole-Pole, Pole-Dipole and Dipole-Dipole. For the case studies shown in this contribution three different array geometries were used, Wenner,

```
      Wenner                    Wenner-Schlumberger                Dipole-Dipole
   ┌─────────┐                 ┌─────────────────┐              ┌──┐       ┌──┐
  ↓ -a- ↓-a- ↓ -a- ↓          ↓ -na- ↓-a-↓ -na- ↓              ↓-a-↓ -na- ↓-a-↓
  A     M    N     B           A      M   N      B              A   B      M   N
```

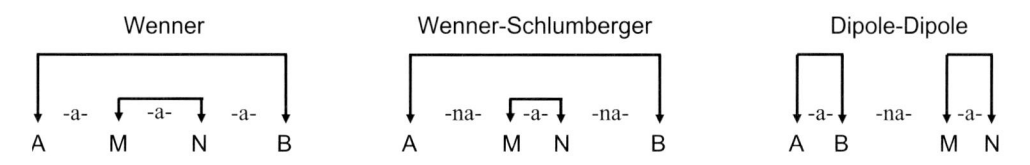

Fig. 2. Scheme of array geometries applied in this study.

Wenner-Schlumberger and Dipole-Dipole (Fig. 2). The corresponding geometric factors are given in geophysical textbooks (e.g. TELFORD et al. 1990, REYNOLDS 1997). For the Wenner array, the two outermost electrodes (A and B) are used as current electrodes while the two electrodes in between are the potential electrodes (cf. Fig. 2). Potential electrode spacing increases as current electrode spacing increases at the same distance. The Wenner configuration has a moderate investigation depth and has a good resolution for horizontal structures with vertical changes of resistivity. Since the total number of measurements required is less than for other configurations the time to complete a survey is comparatively short, however, also the obtained information of the subsurface is less than that derived from other arrays. The Wenner-Schlumberger array is a combination of the Wenner array and the Schlumberger array (commonly used for vertical resistivity soundings, see above) with constant potential electrode spacing but logarithmically increased current electrode spacings leading to a better depth resolution compared to the Wenner configuration. The number of measurements is more than for a Wenner survey but less than for a Double-Dipole array. Since the Wenner-Schlumberger configuration is a combination of Wenner and Schlumberger array types, it is useful for horizontal and vertical geomorphological structures and can be the best choice as a compromise between the Wenner and the Dipole-Dipole array.

The Dipole-Dipole array comprises two dipoles formed by the current electrodes on one side and the potential electrodes on the other side. The current and voltage spacing are the same and the spacing between them is an integer multiple (n) of the distance (a) between the current electrodes (A, B) and the potential electrodes (M, N) (cf. Fig. 2). This array type has a better horizontal, but weaker vertical resolution and a shallower investigation depth than the Wenner array. Furthermore, the largest number of readings of all three presented configurations are required to complete a survey. Further details of advantages and disadvantages of the different array geometries are given for instance in DAHLIN & LOKE (1998), OLDENBURG & LI (1999), OLAYINKA & YARAMANCI (2000).

The data processing of the measured sets of apparent resistivities was performed using the software package RES2DINV. This program supports an implementation of the least-squares method based on a quasi-Newton optimisation technique (LOKE & BARKER 1995, 1996). A homogeneous earth model is used as the starting model, which is obtained by calculating the logarithmic average of the measured apparent resistivity values. The optimisation method tries to reduce the difference between the calculated and measured apparent resistivity values by adjusting the resistivity of the model blocks. A measure of this difference is given by the root-mean-square error (RMS). However, the best model from a geomorphological or geological perspective might not be the one with the lowest possible RMS. As for the interpretation of one-dimensional resistivity soundings, it is thus essential to perform the interpretation with consideration of the local geomorphological setting. This enables unrealistic images of the subsurface structure to be excluded. Additionally, topography may be incorporated in the inversion, which may lead to a more comprehensive resistivity image for the end-user (cf. case studies 3 and 4).

Applications and case studies

Case study 1

Motivation: Detection of the thickness of aeolian sediments and depth of the groundwater table

Fig. 3 shows the result of a 2-D resistivity survey using the Wenner-Schlumberger configuration. This array was applied as it was known that the subsurface consists of relatively homogenous aeolian sand of considerable depth. Furthermore, the terrain available for the survey line was limited, thus leading to the decision of performing a Wenner-Schlumberger measurement with 2 m spacing to obtain reasonable vertical and horizontal resolution. The pseudo-section (Fig. 3) of resistivities against depth shows a general decrease in resistivity with depth. The resistivity pattern in the uppermost meters is influenced by the differing soil moisture contents. The highly conductive zone in the bottom part of the resistivity model is expected to represent the ground water. The dashed line was added into Fig. 3 to mark the boundary of the ground water table. The bulge of the contours around station 50 and 52 is assumed to be caused by a nearby well, which was visible more clearly in the results of an additional Dipole-Dipole survey with 0.5 m spacing (not shown). The latter was performed to obtain a better horizontal resolution. The groundwater table was not reached with the resulting penetration depth. However, low resistivity values (around and below 100 Ωm) limited to the location of the well and at 2 m depth indicate ground water in a host material of markedly higher resistivities which corresponds to the aeolian sand of different moisture contents. Direct observation in the well indicated the depth of the water to be at 1.8 m in the well at the survey day.

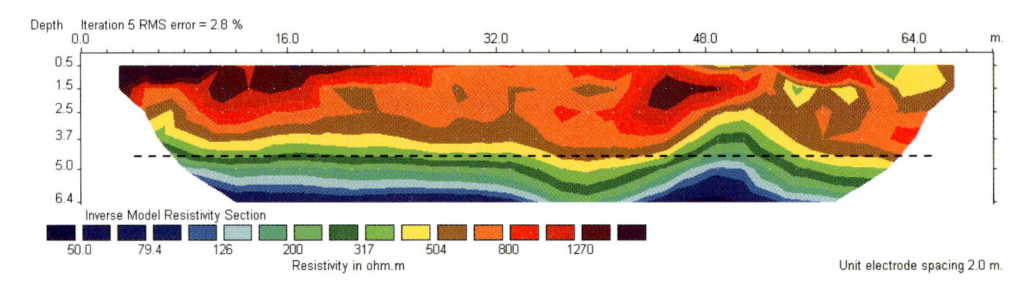

Fig. 3. Resistivity tomogram of a Wenner-Schlumberger survey over aeolian sand.

Case study 2

Motivation: Mapping of the lithology of the near surface unconsolidated sediments and depth to bedrock determination

Fig. 4 shows the result of a 2-D resistivity survey performed to determine the subsurface structure in an area with known great heterogeneity concerning the unconsolidated sediments overlying the bedrock. The bedrock is formed by schist and the overlying sediments are loess loam. Conventional shallow drilling was performed for detailed characterisation of the sediment grain size distribution, however bedrock could not be reached with the light-weight drilling instrument.

The schist and the overlying loess loam do not show marked differences in the resistivity values since the measurements were taken in March with a high moisture saturation in the sediments and the weathered schist. The interpreted depth to bedrock is highlighted with the dashed line. Interpretation of the tomogram is supported by findings of the shallow core drilling which confirmed unconsolidated sediments down to 8 m depth in the survey area. The different moisture contents of the loess loam cause the inhomogeneous resistivity pattern. The near surface small high-resistive anomaly at the left side of the image (around station 40) can be assumed to be caused by an animal den. This example indicates that the interpretation of electrical resistivity tomograms is made difficult in cases of indistinct resistivity changes and overlapping resistivity values of different materials resulting in a non-unique information of the subsurface, if no results from other geophysical methods or drillings are available. Here, at this site, it can be assumed, that the realisation of the survey in summer with drier sediments overlying the schist would have given better contrasts in the subsurface resistivity distribution.

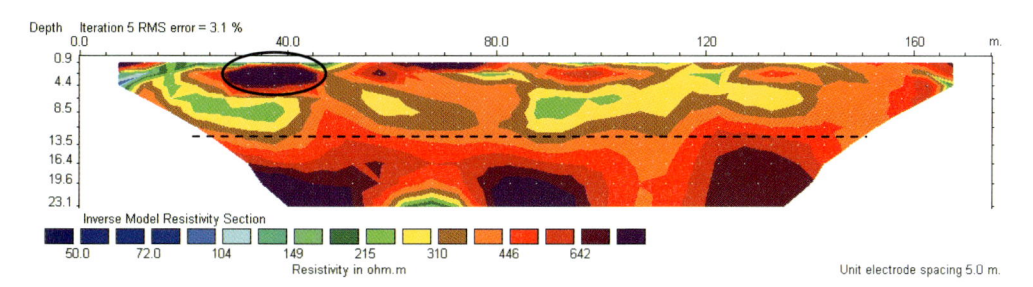

Fig. 4. Resistivity tomogram of a Dipole-Dipole survey performed over weathered schist and overlying loess loam.

Case study 3

Motivation: Detecting permafrost and delineating depth of active layer

A two-dimensional resistivity survey was performed in a periglacial environment in the eastern Swiss Alps close to two perennial snow patches as first indicators of a possible mountain permafrost occurrence. To improve the contact of the steel electrodes to the stony unvegetated ground, sponges soaked in water were used. Topography was included since between station 30 and 60 a depression exists. At this site, one-dimensional geoelectrical soundings were performed in a previous study which already indicated a permafrost occurrence. The results of the two-dimensional Wenner survey confirm the presence of permafrost (cf. blue colors in Fig. 5). A rise of the permafrost table towards the perennial snow patch, which could be expected, is visible in the resistivity image (station 160–170). In the area where the surface has become snow-free only recently the active-layer is thinner as compared to the more central parts of the survey line. This is also evident in the depression between station 45 and 55. The comparatively thick active layer in the central part of the survey could be an evidence of a melting permafrost occurrence at this site.

Resistivity values for permafrost can vary over a wide range (cf. Table 1) depending on the temperature, the ice content and the content of impurities. The resistivities are fairly low, which would usually indicate a low ice content. This pseudosection is interpreted as a permafrost occurrence which consists of a large amount of frozen material of different grain-size, partly ice-rich, rather than a massive ice layer.

Through the one-dimensional sounding similar thickness of the active layer and depth of the permafrost were obtained (KNEISEL 1999). However, since one-dimensional geoelectrical soundings produce only an average resistivity model of the survey area, the lenticular high-resistive areas as shown in the 2-D image could not be deduced. Here, the potential of electrical resistivity tomography for geomorphological investigations compared to one-dimensional soundings shows clearly, as a more realistic image of the subsurface can be derived.

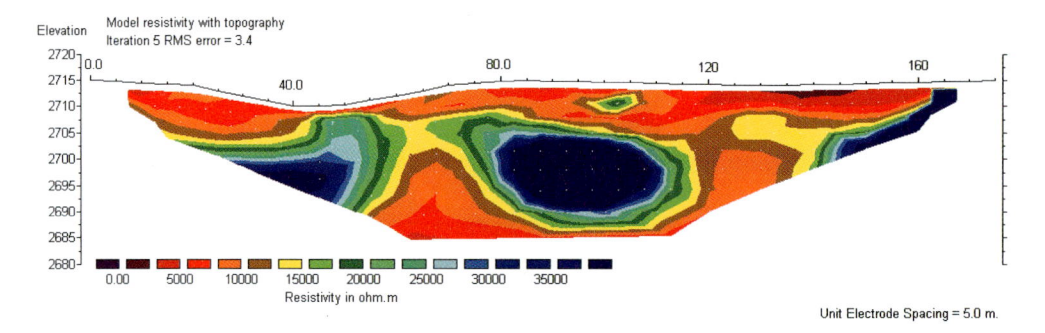

Fig. 5. Resistivity tomogram of a Wenner survey in the alpine periglacial zone with mountain permafrost. (Note the reverse colour scheme for the permafrost case studies).

Case study 4

Motivation: Detecting ground ice/permafrost and delineating depth of scree slope

A Wenner survey was performed along a scree slope in the periglacial belt. Due to the close rockwall and the loose debris only station 0 to 120 were placed at the steep scree slope whereas the rest of the stations are situated at the foot of the slope. Because of high contact resistances of the coarse blocky surface material, sponges soaked in water were used at each electrode to establish sufficient electrical contact to the ground.

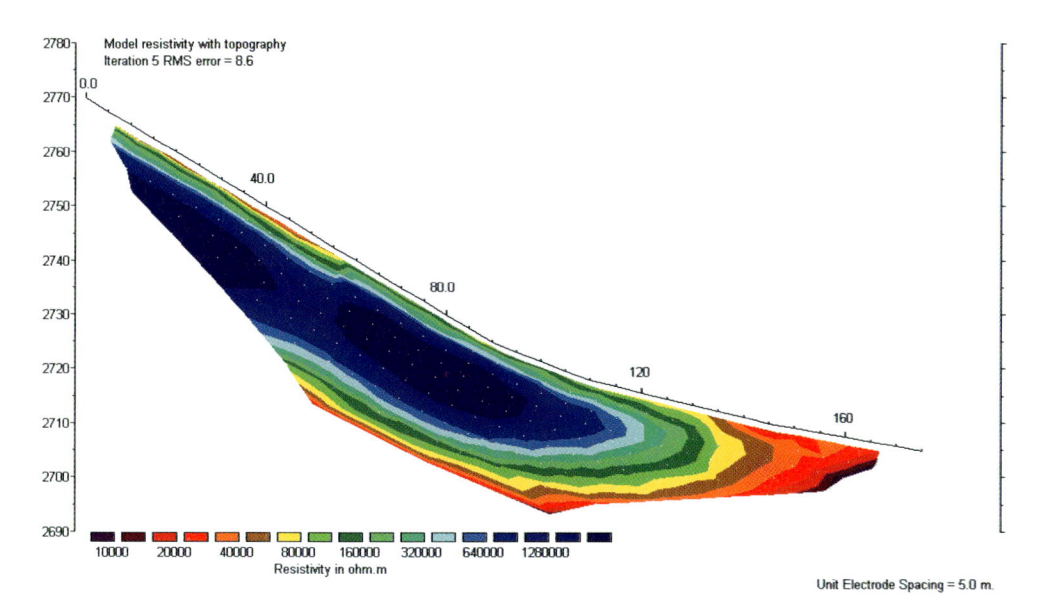

Fig. 6. Resistivity tomogram of a Wenner survey on a scree slope in the alpine periglacial zone with mountain permafrost. (Note the reverse colour scheme for the permafrost case studies).

Massive ground ice of considerable depth could be confirmed in the scree slope which is visible in the high resistivities in the left and central parts of the resistivity image (blue colors in Fig. 6). The range of resistivities with values higher than 1 MΩm allow the assumption that the massive ground ice is of polygenetic origin, consisting of sedimentary ice from a firnification process (typical glacier ice) and congelation ice (typical permafrost ice) from ground freezing (KNEISEL 1999, KNEISEL et al. 2000a for further details).

Depth to bedrock is difficult to delineate from the results of this survey since even the bottom resistivities are still values indicating permafrost. Air-filled voids or cavities which could cause the high resistivities can be excluded. Especially the upper part of the scree slope is creeping through the weight of a person walking uphill or downhill leading to downslope movement of the loose debris of different grain size. Thus, prohibiting cavities to exist. Furthermore, a small debris flow next to the survey line gives geomorphological evidence of active layer slides. The near-linear decrease of resistivity values with depth below about 15 m indicates a continuous decrease in ice content. The active layer appears to be fairly thin compared to the survey in Fig. 5. At the foot of the slope (between station 135 to 165) permafrost cannot be excluded since values higher than 20 kΩm, a figure which is considered to indicate permafrost in non-bedrock areas are to be found.

Case study 5

Motivation: Comparison of Wenner and Dipole-Dipole configurations for surveys on high resistive terrain

To detect isolated permafrost in a scree slope different geophysical techniques were applied during several field trips (KNEISEL & HAUCK, this volume). In order to evaluate the sensitivity of the survey results to the chosen electrode configuration, both Wenner and Dipole-Dipole arrays were tested and compared. The results are illustrated in Fig. 7, both the location and the depth of the permafrost lenses compare well. Note the different depth axes. The lenticular areas with resistivity values as high as 120 kΩm are interpreted as permafrost lenses and they coincide very well with the findings from vertical electrical soundings and from refraction seismics (cf. KNEISEL et al. 2000b). On this vegetated scree slope, although being high-resistive permafrost terrain, the current flow was not too bad compared with other permafrost terrain such as rock glaciers. Therefore, it was appropriate to carry out a Dipole-Dipole survey since first, signal strength was sufficient to obtain reliable results and second, the investigation depth is sufficient for mapping the permafrost lenses.

A comparison of the results of the Wenner array with the results of the Double-Dipole array indeed shows that the latter provide superior horizontal resolution, but a smaller penetration depth as for the Wenner array. In the example in Fig. 7 the results of the Wenner array seem to be sufficient for the overall detection of permafrost presence/absence at this site.

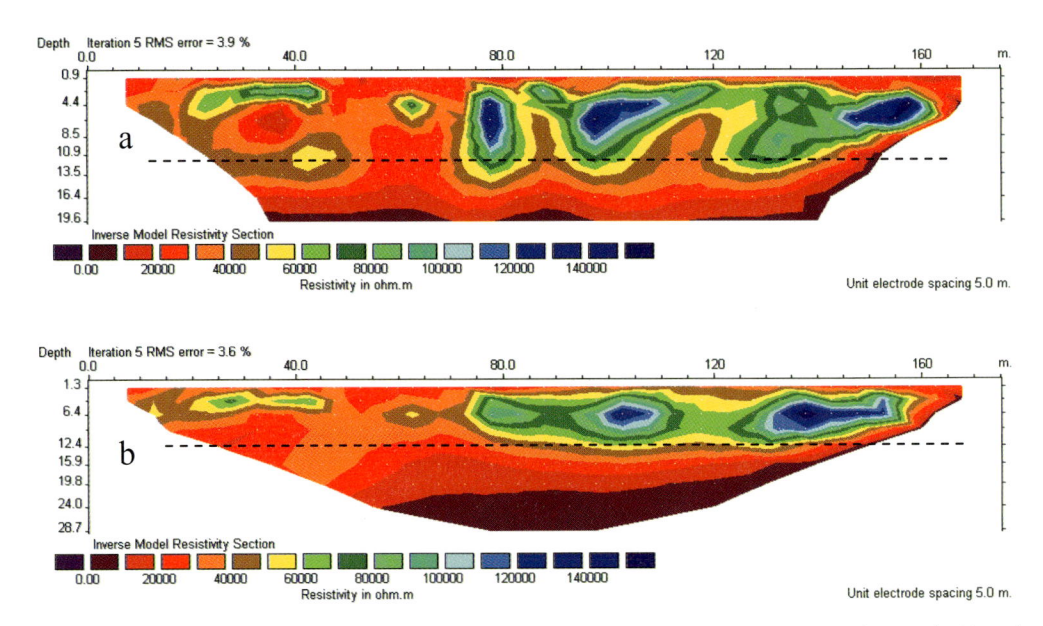

Fig. 7. Comparison of two resistivity tomograms performed at the same location; Dipole-Dipole (a) and Wenner survey (b) on a scree slope with isolated permafrost lenses. (Note the reverse colour scheme for the permafrost case studies).

However, for a detailed characterisation of the location and the extent of the permafrost lenses, the results of the Dipole-Dipole array provide more information.

On more difficult terrain as often associated with mountain permafrost (cf. case studies 3 and 4) choice of an appropriate electrode configuration can be dependent on difficult surface conditions (cf. HAUCK 2001). In this terrain, for large spacings and/or for Dipole-Dipole surveys where the reception voltage values drop quickly with the increasing of the spacing between the injection and reception dipoles the use of the Dipole-Dipole array might be critical. Since the maximum current injected into the ground can be quite low as a result of rough surface conditions with high ground resistance and weak signal strength, the geometrical factors of the electrode configurations may be critical (e.g. TELFORD et al. 1990). Furthermore, two important factors for remote high mountain field work can be the time available for the field measurements (efficiently using good weather conditions) and battery capacity. For these reasons, the Wenner configuration may be the best compromise for very resistive and heterogeneous mountain permafrost terrain especially because it is less sensitive to weak signal strength.

Conclusions

The significance of modern geophysical methods such as two-dimensional resistivity tomography for various environmental studies is beyond controversy, if the limits of data interpretation are considered. 3-D resistivity surveys can even enhance the potential of this method. However, at present 2-D surveys are the best compromise for efficiently obtaining accurate survey results.

The described case studies show that electrical resistivity tomography is a powerful tool also for various geomorphological problems. The resulting pseudosections yield – depending on the array geometry and chosen spacing – to detailed images of the subsurface. For a reliable interpretation of the survey results, knowledge of the geomorphological and geological setting is essential. Furthermore, the following key points can be summarized:

- Depending on the structure of the subsurface, different array geometries can lead to different electrical resistivity tomograms and thus may complicate the interpretation.
- Choice of the appropriate electrode configuration for a field survey has to be determined from case to case. Special characteristics of the different array geometries should be considered, above all the investigation depth and the sensitivity of the array to vertical and horizontal changes in the subsurface resistivity distribution.
- The Dipole-Dipole array provides superior lateral resolution and will often be the first choice for geomorphological applications with expected lateral heterogeneity.
- However, on rough surface conditions with high ground resistance and weak signal strength which frequently occur in periglacial terrain, the application of Dipole-Dipole surveys might be critical.
- For this reason, the Wenner configuration may be the best choice for very resistive and heterogeneous terrain because it is less sensitive to weak signal strength.
- Application of two different array geometries at the same survey site such as Wenner and Dipole-Dipole enables a more accurate and reliable interpretation of the subsurface.

- In difficult cases and to avoid ambiguous results two-dimensional electrical resistivity surveys should be used in conjunction with other geophysical techniques such as refraction seismics or ground penetrating radar surveys as they provide complementary information about the subsurface.
- For detailed geomorphological/sedimentological studies a combination of electrical resistivity tomography with conventional drillings should be applied.

Acknowledgements

Many thanks to S. Friedel, L. King and A. Hoerdt for valuable comments on an earlier version of the manuscript.

References

BARKER, R. & MOORE, J. (1998): The application of time-lapse electrical tomography in groundwater studies. – The Leading Edge **17**: 1454–1458.

BENDERITTER, Y. & SCHOTT, J. J. (1999): Short time variation of the resistivity in an unsaturated soil: the relationship with rainfall. – Europ. Journ. Environm. Engin. Geophys. **4**: 15–35.

CHAMBERS, J., OGILVY, R., KURAS, O., CRIPPS, J. C. & MELDRUM, P. (2002): 3D electrical imaging of known targets at a controlled environmental test site. – Environm. Geol. **41**: 690–704.

DAHLIN, T. (1996): 2D resistivity surveying for environmental and engineering applications. – First Break **14**: 275–284.

DAHLIN, T. & LOKE, M. H. (1998): Resolution of 2D Wenner resistivity imaging as assessed by numerical modeling. – Journ. Appl. Geophys. **38**: 237–249.

DANOWSKI, G. & YARAMANCI, U. (1999): Estimation of water content and porosity using combined radar and geoelectrical measurements. – Europ. Journ. Environm. Engin. Geophys. **4**: 71–85.

GRIFFITHS, D. H. & BARKER, R. D. (1993): Two-dimensional resistivity imaging and modelling in areas of complex geology. – Journ. Appl. Geophys. **29**: 211–226.

HAUCK, C. (2001): Geophysical methods for detecting permafrost in high mountains. – PhD-thesis, ETH Zürich, Mitt. Versuchsanst. Wasserbau, Hydrologie u. Glaziologie der ETH Zürich **171**, 204 pp.

HAUCK C. & VONDER MÜHLL, D. (2003): Evaluation of geophysical techniques for application in mountain permafrost studies. – Z.Geomorph. N. F., Suppl.-Bd. **132**: 161–190.

JOHANSSON, S. & DAHLIN, T. (1996): Seepage monitoring in an earth embankment dam by repeated resistivity measurements. – Europ. Journ. Engin. Geophys. **1**: 229–247.

KNEISEL, C. (1999): Permafrost in Gletschervorfeldern – Eine vergleichende Untersuchung in den Ostschweizer Alpen und Nordschweden. – Trierer Geogr. Stud. **22**, 156 pp.

KNEISEL, C. (2002): Anwendung geoelektrischer Methoden in der Geomorphologie – dargestellt anhand verschiedener Fallbeispiele. – Trierer Geogr. Stud. **25**: 7–20.

KNEISEL, C., HAEBERLI, W. & BAUMHAUER, R. (2000a): Comparison of spatial modelling and field evidence of glacier/permafrost relations in an Alpine permafrost environment. – Ann. Glaciol. **31**: 269–274.

KNEISEL, C., HAUCK, C. & VONDER MÜHLL, D. (2000b): Permafrost below the timberline confirmed and characterized by geoelectrical resistivity measurements, Bever Valley, eastern Swiss Alps. – Permafrost Periglac. Process. **11**: 295–304.

KNEISEL, C. & HAUCK, C. (2003): Multi-method geophysical investigation of a sporadic permafrost occurrence. – Z. Geomorph. N. F., Suppl.-Bd. **132**: 143–157.

KOSZINSKI, S. & WENDROTH, O. (2001): Field-scale variability of soil physical properties and specific electrical resistivity. – 26th General Assembly of EGS, Nice, Geophys. Res. Abstracts **3**: 2128.

LOKE, M. H. & BARKER, R. D. (1995): Least-squares deconvolution of apparent resistivity. – Geophysics **60**: 1682–1690.

LOKE, M. H. & BARKER, R. D. (1996): Rapid least-squares inversion of apparent resistivity pseudosections using a quasi-Newton method. – Geophys. Prospect. **44**: 131–152.

LÜCK, E., EISENREICH, M., DOMSCH, H. & BLUMENSTEIN, O. (2000): Geophysik für Landwirtschaft und Bodenkunde. – Stoffdynamik in Geosystemen, Band 4, 167 pp., Universität Potsdam.

MOORE, J., BARKER, R. & HERBERT, A. (2000): Time-lapse electrical imaging applied to groundwater pump tests and landfill drainage. – Proc. 6th EEGS conf., Ext. Abstracts, P-EL02, 2 pp.

OGILVY, R., MELDRUM, P. & CHAMBERS, J. (1999): Imaging of industrial waste deposits and buried quarry geometry by 3-D resistivity tomography. – Europ. Journ. Environm. Engin. Geophys. **3**: 103–113.

OLDENBURG, D. W. & LI, Y. G. (1999): Estimating depth of investigation in dc resistivity and IP surveys. – Geophysics **64**: 403–416.

OLAYINKA, A. & YARAMANCI, U. (2000): Assessment of the reliability of 2D inversion of apparent resistivity data. – Geophys. Prospect. **48**: 293–316.

PANISSOD, C., MICHOT, D., BENDERITTER, Y. & TABBAGH, A. (2001): On the effectiveness of 2D electrical inversion results: an agricultural case study. – Geophys. Prospect. **49**: 570–576.

REYNOLDS, J. M. (1997): An Introduction to applied and environmental geophysics. – Chichester.

TELFORD, W. M., GELDART, L. P. & SHERIFF, R. E. (1990): Applied geophysics. – 2nd edition, Cambridge University Press.

Address of the author: Christof Kneisel, Institut für Geographie, Universität Würzburg, Am Hubland, D-97074 Würzburg.

Moisture distribution in rockwalls derived from 2D-resistivity measurements

Oliver Sass, Augsburg

with 11 figures and 3 tables

Summary. In this paper the results of a trial application of 2D-geoelectrical soundings ("ERT") are presented. The aim of the field work was to examine the small-scale moisture distribution in rockwalls and its temporal variations, and thereby to establish the presence or lack of water for weathering processes. To achieve this object, the 50 electrodes of a GeoTom ERT unit were replaced by steel nails sunk into the rock with an electrode spacing of 0.04 m (total length of the survey line: 1.96 m).

The measured resistivity values range from 100 Ωm to more than 100 kΩm, which are equivalent to a moisture content of 7.5% to < 0.1%. The highest moisture content for each survey site corresponds well with the pore volume of the rock investigated, while the comparison with moisture data derived from dried and weighed rock pieces collected shows only poor correlation. The moisture distribution in the rock investigated is extremely variable. In most locations, the deeper parts of the rock are predominantly saturated, while the outer 10–15 cm are more or less dried out. Therefore, shallow frost cycles with a penetration depth of less than 10 cm may be insignificant for frost weathering in large parts of the rock. The most important factor controlling the moisture content seems to be the micro-topography of the sites. Joints, ledges or less inclined parts of the rock allow water to soak in and cause moister conditions in their sphere of influence. In addition, south-facing sites are mostly drier than north-facing sites, though micro-topographic differences may cause exceptions to the rule.

The spreading of moisture within the rock seems to take place mainly from the damper inside to the outer parts. The weather conditions in the weeks before measurement are of greater importance for the moisture content than the weather on the day of measurement itself.

Zusammenfassung. 2-D-geoelektrische Messungen zur Ermittlung der Feuchteverteilung in Felswänden. In diesem Aufsatz werden die ersten Ergebnisse einer neuartigen Anwendung der Geoelektrischen Tomographie (ERT) vorgestellt. Das Ziel der Untersuchungen war es, die kleinräumige Feuchteverteilung in einer Felswand in ihrer zeitlichen Variabilität zu erfassen, und damit festzustellen, welche Wassermengen für Verwitterungsprozesse zur Verfügung stehen. Zu diesem Zweck wurden die 50 Elektroden eines ERT-Geräts des Typs GeoTom durch Stahlnägel ersetzt, die im Abstand von 4 cm in den Fels eingebohrt wurden (Gesamtlänge der Profile: 1,96 m).

Die ermittelten Widerstandswerte reichen von 100 Ωm bis mehr als 100 kΩm, was mit einem Feuchtegehalt von 7.5% – <0.1% gleichgesetzt werden kann. Die höchsten ermittelten Feuchtegehalte jedes Standorts stimmen gut mit dem nutzbaren Porenvolumen der untersuchten Gesteine überein. Der Vergleich mit Feuchtegehalten, die durch Trocknen und Wiegen ermittelt wurden, zeigt dagegen eine nur mäßige Übereinstimmung, da die Feuchteverteilung in den untersuchten Gesteinen extrem inhomogen ist. An den meisten Standorten ist das Felsinnere meist wassergesättigt, während die äußeren 10–15 cm mehr oder weniger stark ausgetrocknet sind. "Flache" Frostereignisse mit einer Eindringtiefe von weniger als 10 cm sind daher vermutlich nicht in der Lage, zur Frostverwitterung beizutragen.

Der wichtigste steuernde Faktor für den Feuchtegehalt ist vermutlich die Mikrotopographie. Entlang von Klüften, Absätzen und flacheren Felsbereichen kann Wasser einsickern und zu feuchteren Verhältnissen führen. Daneben sind südexponierte Standorte meist trockener als nordexponierte, wobei jedoch die topographischen Verhältnisse diese Regel außer Kraft setzen können. Die Ausbreitung der Feuchte im Fels scheint hauptsächlich vom feuchteren Felsinneren zur Oberfläche hin zu erfolgen. Die Witterung in den Wochen vor der Messung scheint bedeutender für den Feuchtegehalt des Felsens zu sein als die Wetterbedingungen am Meßtermin selbst.

Introduction

Frost weathering is responsible for the major part of rock disintegration in Arctic and alpine areas. However, the process itself isn't understood in detail. For several decades, the number of freeze-thaw cycles was considered to play the main role in rock disintegration. However, there is an apparent discrepancy between the microsites experiencing a high number of frost cycles (generally south-facing sites), and those said to experience the most intense frost weathering on north-facing rockwalls (HÖLLERMANN 1964, COUTARD & FRANCOU 1989, SASS 1998). For this reason, several authors emphasize that the moisture content of the rock may be just as important as the temperature regime itself. MATSUOKA (1991) established the surprising result, that of four investigated rockwalls, the site in the humid Japanese highlands experienced higher weathering rates than the cold but dry Antarctic site. FAHEY & LEFEBURE (1988) found the most intense rockfall to appear in periods of visible wetness along a rockwall surface. The contradiction between the main weathering factors "frost temperature" and "moisture availability" is especially apparent at snowfield sites, as THORN (1979) points out: "Moisture rich microsites lack adequate freezing intensity, while adequately frozen sites lack moisture" (p. 211).

Because of the 9% expansion of water during freezing, it was assumed for a long time that a pore saturation of over 90% was the precondition for frost weathering. However, there is an apparent contradiction between this presumption and the conditions in most areas investigated, as WHITE (1976, S. 6) points out: "In how many mountain ranges [...] will bedrock fortuitously ever become >90% water-saturated from melting snow or rain and then undergo rapid freezing to crack the rock?" In a later publication a "critical degree of saturation" well below 90% was postulated (McGREEVY & WHALLEY 1985). This does not take into consideration that pore properties and pore distribution in natural rockwalls may vary over a very small area. WALDER & HALLETT (1986) delineate the importance of water movement from small pores to larger cracks. The importance of water for weathering is also emphasized by various investigations concerning rock disintegration by wetting and drying without the contribution of frost (e.g. HALL & HALL 1996).

Data on the moisture content of rock in different locations are difficult to obtain: "Studies […] add to a substantial and growing body of data on rock temperatures in cold environments. Unfortunately, complementary information on rock moisture conditions is still scarce, probably because of the practical difficulties of measurement" (McGREEVY & WHALLEY 1987, p. 358). Moisture content may be determined by simply drying and weighing of pieces of rock, a method which may be helpful to determine the spatial moisture distribution, but only roughly (e.g. HALL 1986, 1991). In addition, the measurements can usually neither be repeated nor reproduced. By means of resistivity records or even more sophisticated techniques like TDR, FDR or microwaves, the temporal variability of rock moisture and its relation to weather conditions can be determined quite well (SASS 1998, SALVE 1998). However, all of these methods give an average reading over a fixed, usually very small part of the rock. Water that is concentrated along joints and bedding planes cannot be detected and may lead to over- or under-estimation of the actual moisture content, depending on the exact location of the probe.

To overcome these limitations and to fill a gap between the various existing methods, the application of ERT (**E**arth **R**esistivity **T**omography) seems to be a promising approach. (The term "tomography" is somewhat misleading, since it is sometimes confined to three-dimensional studies.) Up to now, geomorphologists used geoelectrical soundings chiefly for the detection of permafrost (e.g. KING 1982, WAGNER 1996, EVIN et al. 1997, KNEISEL 2000, HAUCK 2001). The aim of the present study was to create 2D-sections of the moisture distribution of different rock under different weather conditions, in order to find out (1) where the rock is jointed and, thus, where water can percolate, (2) which amounts of water are concentrated along these joints under various conditions, and (3) whether the rock moisture increases or diminishes with bedrock depth.

Methodology

In recent years, 2-D resistivity measurements have become a common tool in geophysics. Resistivity measurements are generally carried out by applying current into the ground through two "current electrodes" and measuring the resulting voltage difference at two "potential electrodes". From the current and voltage values, an apparent resistivity value is calculated (this means the resistivity of a homogenous half-space without geological structures). 2D-surveys are using 25 or more electrodes connected by a multi-core cable. The relevant four electrodes for each single measurement are automatically selected by a switching unit controlled by a microcomputer (LOKE 1999). To determine the true subsurface resistivity in different zones or layers, an "inversion" of the measured apparent resistivity values (generally a total of some 100 single values) must be carried out using a computer program. The result gives information on spatial averages of rock resistivities in a 2D-section. There might be also 3D-effects (structures above or below the survey line) affecting the data. This effect can be investigated by measuring a "cross-section", which remains a possible future topic of investigation.

Since the electrical resistivity of massive or fissured rock is almost exclusively dependent upon its moisture content (LOKE 1999), resistivity measurements allow a good estimation of moisture distribution. In textbooks like LEHNERT & ROTHE (1962) or KNÖDEL et al. (1997) resistivity values for a number of geological units are given. These values are valid for spacious stratigraphical units and usually cover a range of several orders of magnitude. For the present investigation, small-scale values of the investigated rocks had to be derived from laboratory measurements. Six square-formed specimens of different limestones and dolomites were first adjusted to defined moisture contents, then the resistivity was measured (SASS 1998). The result was a calibration curve (Fig. 1), which allows to calculate the small-scale moisture content of the rock from its resistivity. During the laboratory measurements, no significant difference of the moisture/resistivity conversion between the different rock types was considered. However, the two investigated sandstones (see chapter "study sites") have not yet been fully incorporated into the laboratory measurements. Single measurements on sandstones indicate that the application of the limestone calibration curve is possible.

The accuracy of the method is dependent upon the accuracy of the calibration curve. The resolution of the resistivity measurement itself is comparatively high, the possible error is caused mainly by the conversion to moisture values. A standard error of about 0,12% moisture content by weight has to be considered.

The moisture content is strongly determined by the pore space available. The total amount of water is much higher where rock is fractured. Under moist conditions, these

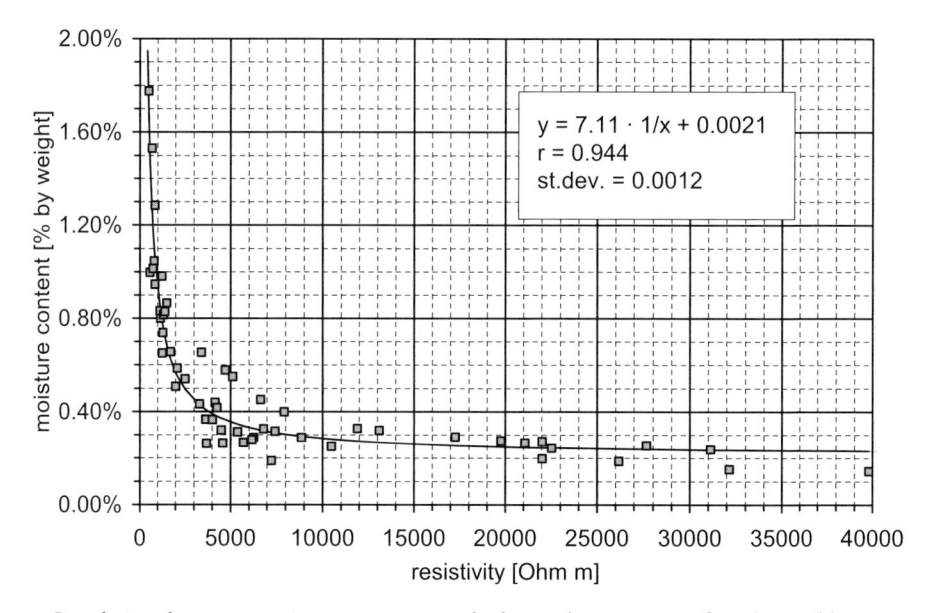

Fig. 1. Correlation between moisture content and electrical resistivity of rocks. Calibration curve derived from laboratory measurements (SASS 1998, modified).

parts become visible as zones of lower resistivity. Inversely, if fractures completely dry out, lower conductivity in the fractured parts seems to result, because an air-filled joint acts as a seperating interstice. The occurence of this depends upon the width of the joints. Calculated conductivity values which exceed the maximum value of saturated massive rock, must be caused by joints that are partly or fully filled by water. Clay minerals along a crack may have a similar effect due to their comparetively high conductivity. However, the investigated limestones and sandstones in this study contain almost no clay minerals.

A further parameter controlling resistivity is temperature. As laboratory measurements show, the resistivity of rock generally decreases with rising temperatures. In the temperatur range between 0° and 40 °C, the conductivity of (pore-) water increases by 2.2% per °C (WEAST et al. 1989). When results from two different days with different temperature are checked against each other, this conductivity difference has to be taken into account. Furthermore, the temperature distribution in the rock investigated may be irregular, especially in south-facing microsites under sunny conditions, where the surface and the outermost centimetres of the rock may be several degrees warmer. Lower resistivity near the surface may therefore be a temperature induced effect.

Even more problematic for the interpretation of the data is the content of soluble salts in the pore water. Laboratory experiments show that in limestones, the same resistivity occurs regardless of whether distilled water, tap water or even a saturated $CaCO_3$ solution is used for wetting the rock (SASS 1998). However, when salts of higher solubility such as NaCl are added, conductivity rises by a factor of 10 or even 100. To check the validity of the calibration curve, the rock types investigated were additionally lab-examined by X-ray diffractrometry. Since neither NaCl nor $CaSO_4$ were detected, the problem of soluble salts probably can be ignored.

Instrumentation

A GeoTom-2D system equipped with multicore cables for 50 electrodes was used. The connection to the rock was established by steel nails or screws, 5 mm in diameter, which were driven into 4 mm-boreholes, 10 mm deep, 40 mm apart. Thus, the total extent of the survey line was about 2 m (1.96 m). Fig. 2 shows the arrangement at the "Kochel north" site.

The distances between the electrodes and the depth of the electrode boreholes indicate that deviations exist from the theoretical approach of point sources, which might lead to a falsification of the results especially in the uppermost layer of a section. The measuring error can be estimated by employing electrical images (e.g. TELFORD et al. 1990), using the simplifying assumtion that the electrodes are point sources in 1 cm depth. For an electrode spacing of 4 cm, the deviation from the ideal case of point sources at the surface is about 5% in the uppermost sequence. This means that the moisture content of the rock surface may be slightly over-estimated. However, evidence for this deviation could not be established in any of the profiles.

However, some minor technical adaptations of the equipment were necessary:
– When using a current of 0.05 mA (which is common for most instruments), the voltage generated in some parts of the survey profile may be too high. This is because of the

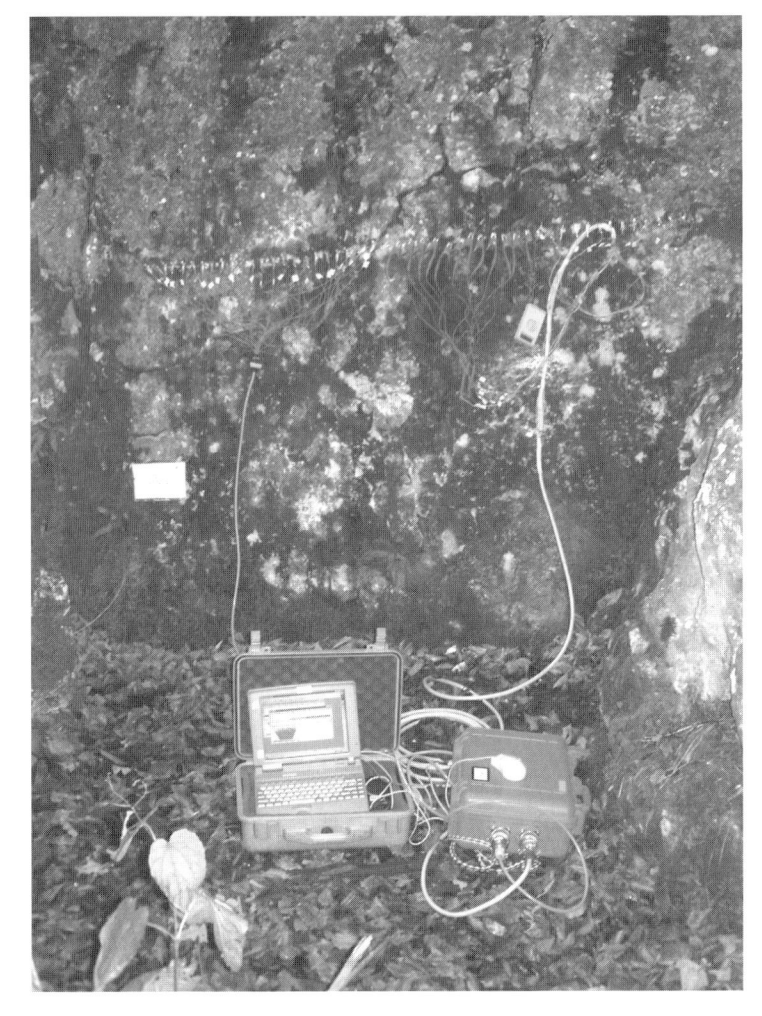

Fig. 2. Arrangement at the "Kochel north" site (11-2001, first attempt with unshielded cables).

extremely high resistivity of dry rock, which may act as a non-conductor when it contains virtually no water. So the current should be as low as possible. In the present investigation, some hardware and software adaptations were made to reduce the amperage to 0.005 mA.
- The maximum voltage to be measured by the GeoTom unit was 60 mV, which proved to be not sufficient. This limitation was overcome by reducing the internal amplification factor of the device by factor 4. Thus, a maximum voltage of 240 mV could be registered. To avoid damage by high voltage within the measuring device, the instrument was earthed by a central electrode stuck into the ground at some distance.
- During high voltage generation, inaccurate or scattered readings may appear due to induction between cables. Shielded multicore cables counteract these problems.

Furthermore, attention should be paid to the contact between electrodes (=nails) and rock. A drop of water was sufficient to reduce the transitional resistivity. Some authors recommended sponges soaked with salt water to establish contact to rocky ground on rock glaciers (e.g. Hauck 2001). In the present small-scale investigation, it is probably not advisable to use salt water in order not to falsify the results.

The measurements were carried out by means of a "Wenner array" and a "dipole-dipole array". In a Wenner array, the distance between the four electrodes (first current electrode C1, first potential electrode P1, second potential electrode P2, second current electrode C2) used for a single reading is kept constant. For the first reading, the electrodes 1, 2, 3 and 4 are used, then 2, 3, 4 and 5 and so on until 47, 48, 49, 50 when using 50 electrodes. Then the second sequence of measurements is done with a spacing of two electrodes. This means that the electrodes no. 1, 3, 5, 7 are used, then 2, 4, 6, 8 and so on. The last possible sequence is carried out with a unit distance of 16 electrodes (1, 17, 33, 49 and 2, 18, 34, 50).

For the dipole-dipole array, the measurement starts with a distance of "1" between the C1-C2 and also the P1-P2 electrode. In each measurement sequence, the distance between these two dipols is increased by one. This array provides a better resolution for vertical structures. However, the Wenner measurements turned out to be much faster in the field. The results were plausible and seemed to produce less scattering caused by varying coupling conditions. For this reason, Wenner arrays were used in all experiments. The Wenner array provides a good depth resolution in the central parts of the profile, while the sensitivity for vertical structures is comparatively low (Knödel et al. 1997, Loke 1999).

During post-processing, some implausible individual readings were eliminated by linear interpolation between the adjacent cells. Such scattered values are probably caused by problems with the electrode coupling to the dry rock. In each profile, only 4 to 8 of 392 readings were edited in this way.

In relation to the extent of the survey lines, most sites show a pronounced topography. Uneven parts of the rock (which may seem to be insignificant at first sight) form steep "hills" and "valleys" in their profile. Thus, consideration of the topography is essential. The profiles were determined by measuring the vertical distances with a long, straight wooden lath.

The inversion was calculated by using the scientific standard program Res2Dinv provided by M. H. Loke (Malaysia). Res2Dinv provides a resistivity model whose response fits the measured data within a certain reliability. The quality of the model fit is given in units of RMS error. Some assorted settings are given in Table 1. These settings turned out to produce the most plausible results with a comparatively low RMS error. The fine mesh and the robust data constrain counteract overdrive effects caused by the high contrasts in resistivity and coupling conditions.

Table 1. Inversion settings.

Initial damping factor	0.15	Data constrain	robust
Minimum damping factor	0.03	Effect of side blocks	reduced
Number of iterations	6	Nodes between adjacent electrodes	4
Increase in thickness of layers	25%	Type of mesh	finest
Increase of damping factor with depth	1.2	Topography modeling	yes

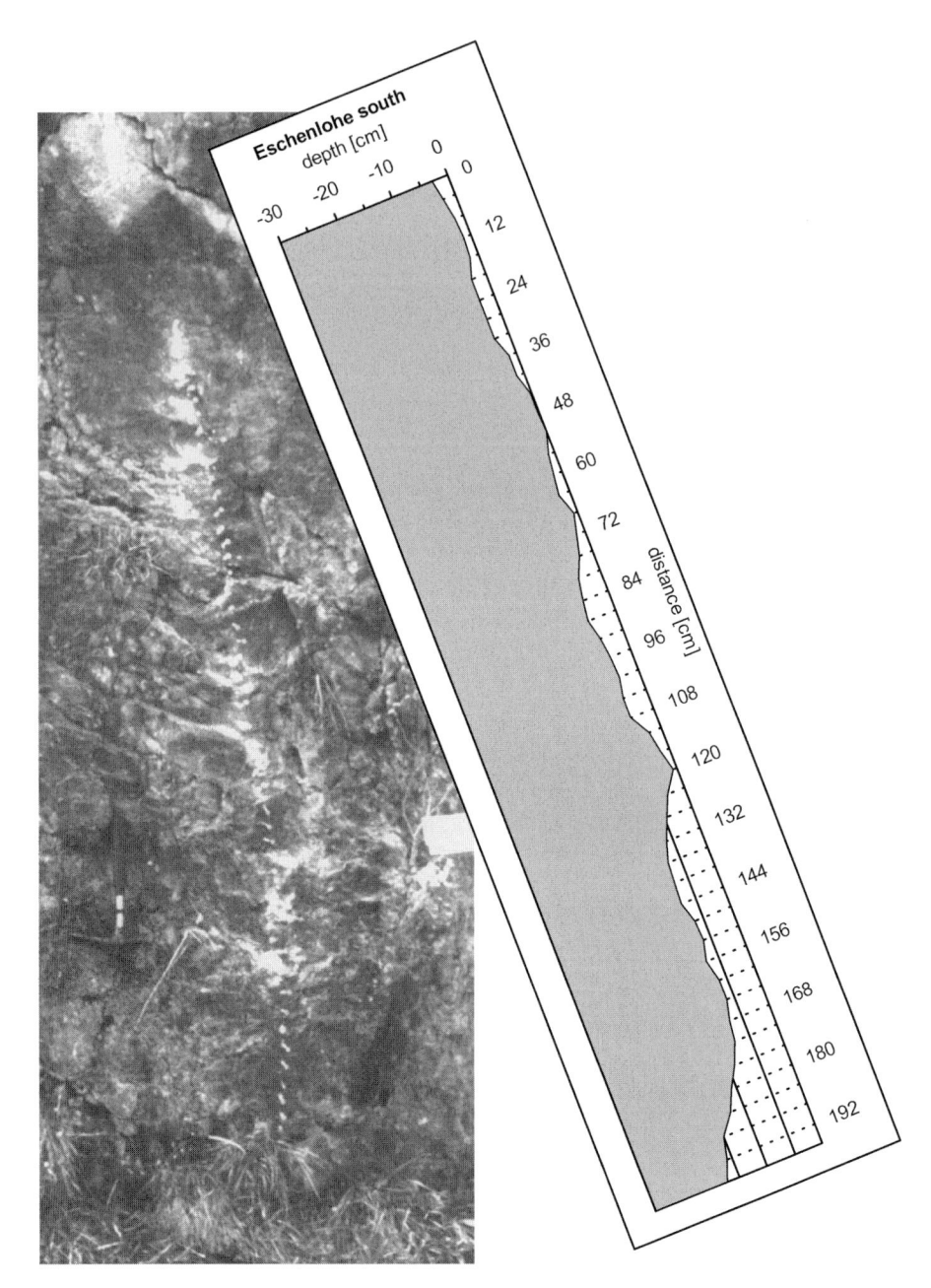

Fig. 3. Topography at the site Eschenlohe south.

Study sites

In this preliminary investigation, three sites at the northern edge of the Bavarian Alps were chosen. The areas cover different geological settings and are easily accessible. The locations are shown in Fig. 4. After testing the suitability of the method outlined above, it is intended to establish two additional study sites in a high alpine environment.

At the northern edge of the Bavarian Alps, several tectonic units were compressed and overthrust to more or less narrow, E-W extending bands. The sandstones of the Bad Heilbrunn and Bichl sites (both former quarries) are related to the Helvetikum zone, which forms small but scenically eyecatching outcrops in the alpine foreland. The "Assilinensandstein" is a grey, sometimes yellowish brown, medium-grained sandstone, while the "Stallauer Grünsandstein" is a more fine-grained, noticeably green, glauconitic sandstone.

The "Wettersteinkalk" and the "Hauptdolomit" are related to the calcareous alpine zone (Oberostalpin). These two rock series form most of the summits in the area of investigation. The Wettersteinkalk is a massive, light-grey, very pure limestone. At the Kochel site, a narrow outcrop forms some small but steep rockwalls. The Hauptdolomit is an intensely fractured, light-grey dolomite. At the Eschenlohe site, it builds up the major parts of the surrounding valley sides. Near the valley bottom, there are some friable rock outcrops in trenches and steep slopes. The geological settings of the sites studied are shown in Table 2.

The joint density was determined by simply counting the visible joints intersecting the sections surveyed. The moisture holding capacity was measured by immersing samples of rock in water for 48 hours under normal atmospheric pressure. The total pore space may be slightly larger than actually measured. However, cavities which are not accessible to water after two days of immersion, are probably not of great importance for the percolation of water under natural conditions. At every site, three samples of 200–500 g were taken. This method can only give a rough outline of the pore space of a large rock volume, for larger cracks are certainly not represented in the relatively small samples.

Before the first day of measurement (20-04-02), there had been several weeks of very dry weather. On 20-04-02 (while measuring), it rained heavily all day. During the days between

Table 2. Geological settings of the study sites.

site	aspect	rock type	local name	joint density	moisture holding capacity (min/mean/max)		
Eschenlohe	north	dolomite	Hauptdolomit	6 /m	1,4%	2,1%	2,8%
	south-east			6 /m	0,3%	0,6%	0,9%
Kochel	north			4 /m	0,6%	0,8%	1,1%
	south 1	limestone	Wettersteinkalk	3 /m	0,5%	0,8%	1,1%
	south 2			6 /m	0,9%	1,5%	2,2%
Bad Heilbrunn	north-west	sandstone	Assilinen-Sandstein	5 /m	2,7%	3,1%	3,5%
Bichl	south	sandstone	Stallauer Grünsandstein	7 /m	4,4%	5,8%	6,8%

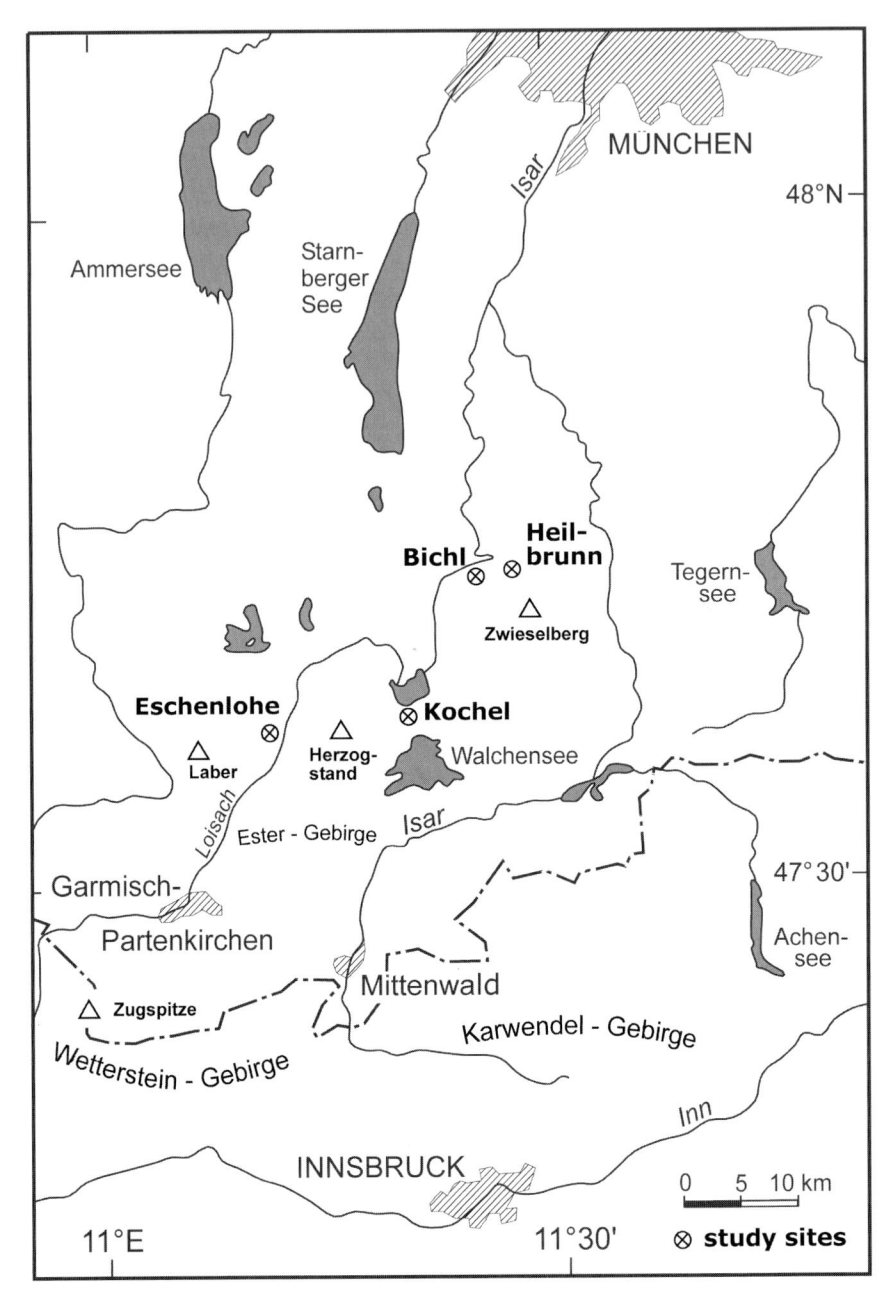

Fig. 4. Location of the study sites.

20-04-02 and 01-05-02, there were several rain showers, but on the second day of measurement it was sunny and dry.

Results

The resistivity values measured range from 0.1 to more than 100 kΩm, which is equal to a moisture content of 7.3% to <0.1% (calculation of moisture contents from the calibration curve given in Fig. 1). For all presented sections (Figs. 5–11), the same linear colour chart was used. The chart was cut off at 25 kΩm (0.24% per weight), all larger values were considered as "completely dry".

From 20-04-02 to 01-05-02 the moisture content generally showed very small differences. At first sight this is surprising, as it rained heavily on 20-04-02, while on 01-05-02 it was sunny. The weather conditions in the weeks or months before measurements are made seem to be of greater importance than the weather on the day of measurement itself. This result is similar for all study sites, with the exeption of "Kochel south 1" and "Kochel south 2", where probably direct radiation caused slightly drier conditions on average. On grounds of the little differences, for all sites only the results of one of the dates are presented.

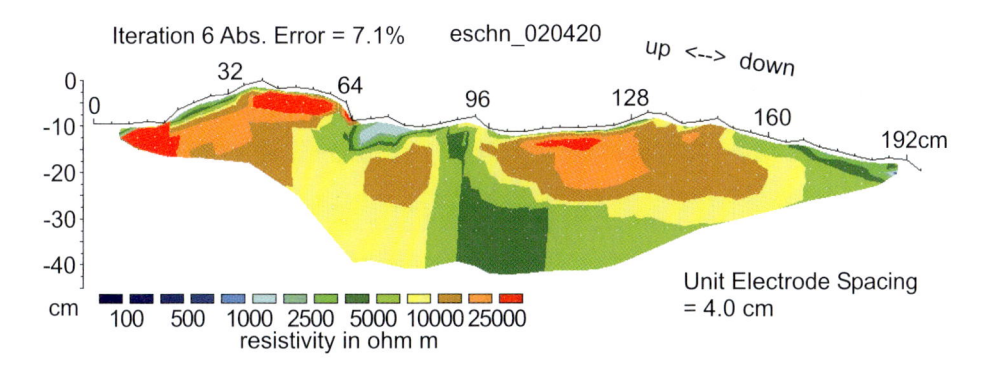

Fig. 5. Inversion result "Eschenlohe N", 20-04-02 (RMS error: 7.1%).

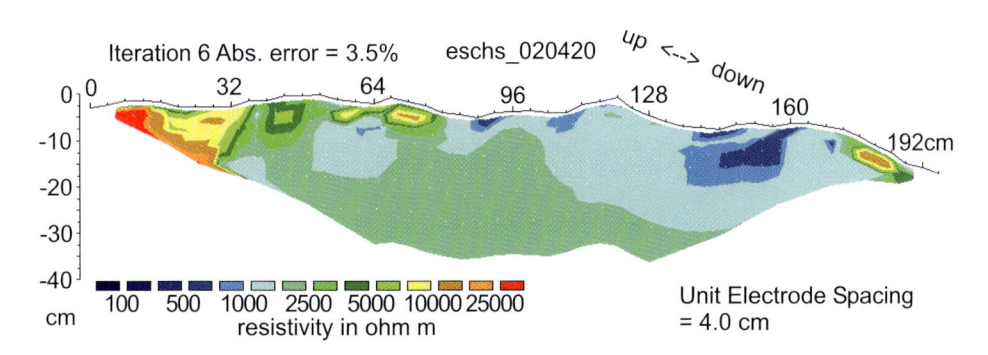

Fig. 6. Inversion result "Eschenlohe S", 20-04-02 (RMS error: 3.5%).

a) Eschenlohe north (fissured dolomite)

The profile stretched nearly from the top down to the foot of a small, almost vertical rock outcrop. The nature of the micro-topography permitted only the upper part (20–40 cm of the profile) to be rained on. Here a slight surface wetting was determined by lower resistivity in these zones (Fig. 5). Below a ledge (40–68 cm), the surface was completely dry, probably because the rock is shielded from direct rainfall. The following wet zone (70–80 cm) coincides with a marked joint that stretches diagonally upwards into the rock. In this position under the overhang, the moisture must have been caused by water flowing out. However, the absolute values (1.5 Ωm / 0.7% moisture content) are fairly low. Downward from 100 cm (profile distance), the rock surface was comparatively dry again. The deeper parts of the rock (20–30 cm in depth) seem to contain more moisture than the surface.

b) Eschenlohe south (fissured dolomite)

The profile (Fig. 6) extends from the top to the bottom of a rock slope with an inclination of about 65°, along the slope cut by a small forest road. In spite of the southerly aspect, this is the position with the highest moisture content among the limestone/dolomite sites. The reason may be the comparatively low inclination, which leads to a larger amount of rainfall on the surface. The uppermost part (0–36 cm) consist of a compact rock face, where obviously only little water can infiltrate. Between 80 and 172 cm, there are several fissures running almost parallel to the surface. These bedding joints seem to act as important paths for the water to infiltrate. Between 32 and 80 cm, water seems to percolate under the drier surface on one of these bedding planes. In some "moisture nodes", a water content of about 1.4% (600 Ωm) was established. This clearly exceeds the pore volume of the (unfissured) rock. Hence, the existence of water-filled joints in these areas is very likely.

On May 1st, the moisture content of some parts of the rock had increased, the water-filled nodes were enlarged (no figure).

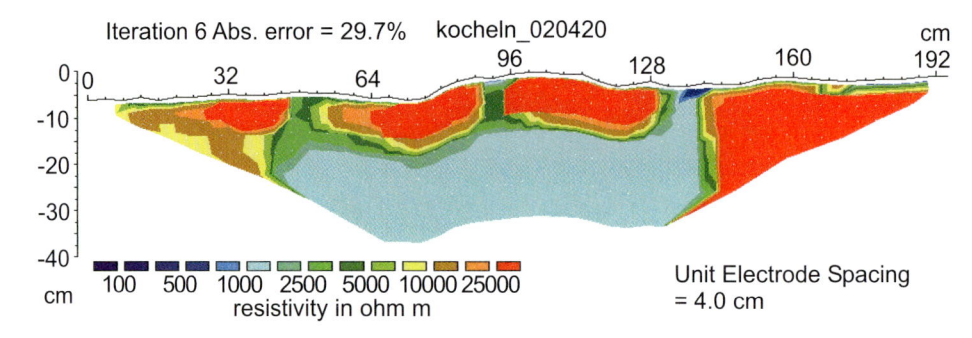

Fig. 7. Inversion result "Kochel N", 20-04-02 (RMS error: 29.7%).

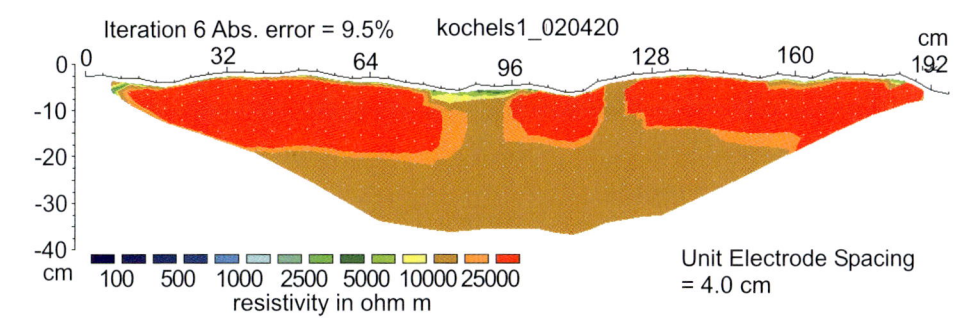

Fig. 8. Inversion result "Kochel S1", 20-04-02 (RMS error: 9.5%).

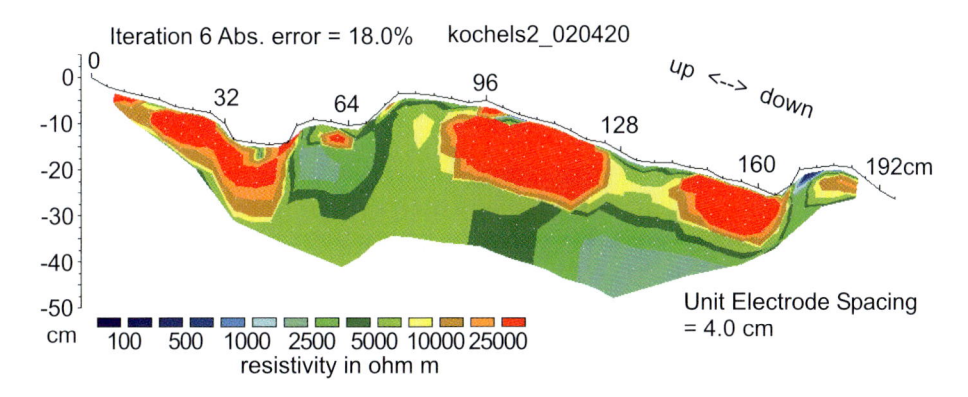

Fig. 9. Inversion result "Kochel S2", 20-04-02 (RMS error: 18.0%).

c) Kochel north (compact limestone)

The profile (Fig. 7) stretches horizontally along a massive rock face with a slope angle of approx. 80°, which is streaked with some fissure lines. In spite of the rather plane topography and the homogenous impression of the site, the measurement as well as the inversion were problematic. On both dates, an RMS error of nearly 30% was found, which was caused by scattered values in several parts of the apparent resistivity section. This may be due to contact problems in the extremely massive (= non-conducting) rock. For this, only a rough interpretation is possible. The surface was extremely dry, apart from the fissure zones, which seem to allow moisture to soak in (or out). The comparatively wet zones between of 92 and 132 cm on the profile exactly match the position of the joints mapped. The deeper parts of the rock were obviously much wetter (about 0.7% moisture content, which means saturation in this position). It is supposed that the water is "inherent moisture", which has not infiltrated from the rock face, but probably from the wooded slope above the rockwall.

An additional measurement was carried out on 03-09-02, after a long period of moist weather conditions. The same general result was found (wet inside of the rock, drier surface, moisture concentration along joints), with markedly reduced coupling problems, probably due to the wetter rock (RMS error = 6.4%, no figure).

d) Kochel south 1 (compact limestone)

This profile (Fig. 8) extended horizontally along a massive, vertical rockwall. The total height of the wall is about 10 m, and the profile was taken 1 m above the ground. It faces due south.

This site was the driest one by far. Inside the rock, the moisture content was no more than 0.26% (15 kΩm). The surface was completely dry down to a depth of 10–15 cm, except for a superficial zone from 76 to 100 cm, where rain water ran down a shallow depression. On 01-05-02, probably due to the sunny weather, the rock was generally even drier, whereas the moisture distribution remained more or less the same. The RMS error of 9.5% was caused mainly by coupling problems of two or three electrodes to the dry rock. However, when the data of these electrodes is deleted and subsequentially interpolated, the inversion result remains more or less the same.

e) Kochel south 2 (friable limestone)

This survey line extended vertically up a small, steep buttress with visible joints and bedding planes. The survey site was comparetively dry, but much more differentiated than Kochel south 1 (Fig. 9). The RMS error is relatively high again.

Infiltration seems to occur along the small, less inclined rock ledges (52–80 cm, 178–180 cm). The vertical zones of the profile are very dry near the surface. This finding is probably due to the better situation for infiltration along joints and on the ledges. The inner parts of the rock are much wetter than at Kochel south 1. However, the moisture content is less than 0.6%, which is far from saturation (1.5%, see Table 2). The wetting of the rock from the surface itself seems negligible, except along the ledges and fissures.

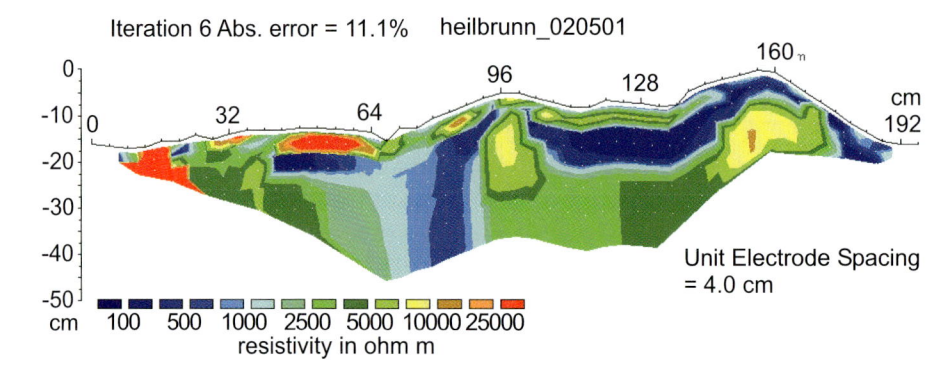

Fig. 10. Inversion result "Heilbrunn", 01-05-02 (RMS error: 11.1%).

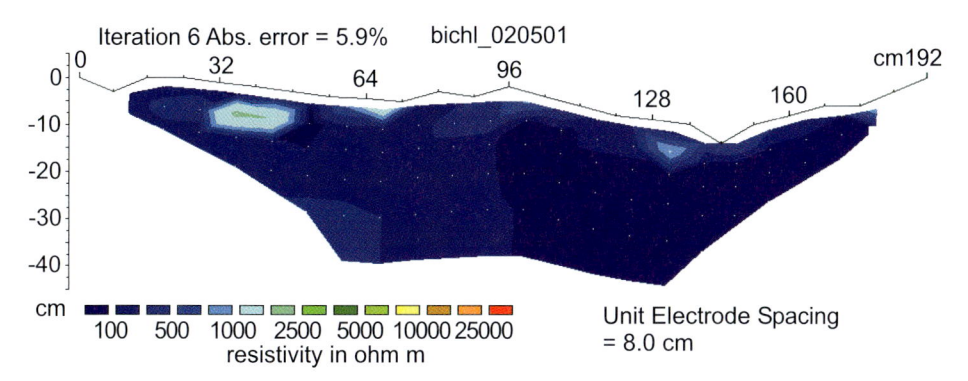

Fig. 11. Inversion result "Bichl", 01-05-02 (RMS error: 5.9%).

f) Bad Heilbrunn (sandstone, northerly aspect)

The profile stretches horizontally along a rather small outcrop (5×5 m) on a steep, forested slope. In sandstone, a more homogenous moisture distribution had been expected, due to the regularity of the intergranular porosity. However, the sections suggest just the opposite. Due to the fewer but wider cracks, the moisture distribution is extremely differentiated (Fig. 10). The wetting of the rock from the surface seems to have been hampered, probably because the small, air-filled, intregranular pores may act like "air-cushions" in which the air is not easy to be replaced by water. The moisture is strongly concentrated along joints where the infiltrating water has more time to soak into the pores. The large, wet zone on the right of the profile corresponds well with the large cracks running above and parallel to the profile line and diagonally down into the rock face (168 cm and 114 cm). In the vicinity of the hypothetical joint system, a water content of up to 5% is reached, which means saturation. By contrast, large areas near the surface are almost completely dry (up to 100 kΩm = 0.1%).

g) Bichl (glauconitic sandstone, southerly aspect)

This profile is situated along a small rock wall on the southern side of an isolated sheepback. Due to the extreme brittleness of the rock, only 25 electrodes spaced 8 cm apart were installed.

The high moisture content of 3.8% (200 Ωm) on the left and 7.3% (100 Ωm) on the right of the profile (Fig. 11) corresponds well with the geological setting. On the left, there is somewhat more massive rock, with a moisture-holding capacity of 4.0 to 4.5%, while the very brittle rock on the right is more porous (6.4 to 7.0%). Due to large fissures, the measured/calculated moisture content may exceed the moisture-holding capacity found in rock samples. Thus, along the entire profile, the moisture content within the rock is near saturation. Evaporation from the surface probably causes the slightly lower values near the surface.

It is highly unlikely that the high moisture content is caused by infiltration from the surface of the almost vertical, south-facing rock. The rainfall probably filters down through the top of the sheepback and seeps out along the rockwall.

Table 3. Comparison between moisture content determined by drying and weighing and moisture content calculated from ERT measurement, 01-05-02.

site	moisture from sample	moisture from ERT (near surface)	correspondence
Eschenlohe N	0.88%	0.3 – 0.8%	nil
Eschenlohe S	0.19%	0.3 – 1.4%	nil
Kochel N	0.36%	0.1 – 0.6%	yes (sample taken near joint)
Kochel S1	0.13%	0.1 – 0.4%	yes
Kochel S2	0.16%	0.1 – 0.6%	yes (sample taken from dry buttress)
Heilbrunn	1.19%	0.2 – 1.5%	? (not clear)
Bichl	2.56%	1.0 – 5.0%	yes

Discussion

As shown by means of ERT measurements, the moisture content in rockwalls is highly varied in both time and space. Therefore, data obtained by taking samples from the rock surface may be misleading or at least inaccurate (Table 3). The micro-topography of the sampling site should be recorded precisely, as to help in the interpretation of the values.

At all of the seven sites, the deeper parts of the rock are wetter than the surface. Direct rainfall seems to contribute little to the moisture content of compact, steep parts of the rock. The results suggest that water is rather infiltrating at joints and ledges (or on less inclined slopes at the top of the rockwall) and slowly disperses in the rock itself, while compact, steep parts are under "arid" conditions with "ascending", outward directed water movement.

The calculated moisture content is in agreement with the pore volume of the rock types investigated. At most of the sites, the deeper parts of the rock ($>10-15$ cm) are saturated, while the surface is more or less dried out. At the Kochel south sites, the moisture content is clearly under the possible maximum. Due to the southerly aspect, the evaporation from the rock surface seems to be greater than the amount of infiltrating water. At the Kochel south 1 site, the dryness is exacerbated by the compact structure of the vertical rock, where very little water can soak in. In contrast to this, the site Eschenlohe south is much wetter than the neighbouring site Eschenlohe north. The higher water content in the south-facing site is probably controlled by the topography and joint orientation. In the north-facing site, only little water can infiltrate along the vertical, bulging rock, while the inherent water is likely to soak out in brittled zones on the left and right of the profile.

The temporal variablity of rock moisture seems to be predominately controlled by the weather conditions in the preceding weeks, while the weather at the time of measuring is less important. Further measurements are likely to produce data concerning the different moisture conditions in different seasons.

Implications for weathering

Frost weathering (and other weathering processes as well) are highly dependent upon the availability of moisture. Weathering is the starting point for debris production and thus for many further geomorphological processes. However, based on different results from different study sites, it is difficult to define universally valid implications in the present state of the investigation.

The susceptibility to weathering in its spatial distribution is likely to be determined by pore volume and joint density of the rock, for these factors essentially control the amount of water available for weathering processes. On a smaller scale, the catchment area for inherent moisture, the orientation of cracks and, last but not least, the aspect are important controlling factors.

The results presented point to a higher moisture content in the inner parts of the rock, while the surface is dried out more or less. As a consequence, shallow frost cycles with a penetration depth of less than 10–15 cm are unlikely to make it to the saturated parts of the rock. Hence, the intensity and the duration of temperatures below freezing point are probably of greater importance for rock disintegration than the number of frost cycles. Shallow freeze-thaw cycles of some hours or minutes in duration, (e.g. caused by cloud shading), may contribute to small-scale desquamation ("micro-gelivation"), but are presumably unable to disintegrate rock by frost wedging ("macro-gelivation"). These cycles may also cause rockfall, but only near brittled zones where water is seeping out, or after long periods of moist weather conditions. This conclusion is particularly important for south-facing rockwalls, where the comparatively high number of shallow, diurnal freeze-thaw cycles (VORNDRAN 1969, COUTARD & FRANCOU 1989) is counteracted by the desiccated rock surface. In north-exposed sites, the frost usually penetrates deeper into the rock, where most of the time there is sufficient water. More intense rockfall in north-exposed rockwalls, as stated by several authors (e.g. HÖLLERMANN 1964, COUTARD & FRANCOU 1989, SASS 1998), may be caused by the outlined moisture distribution. Further interpretations are limited due to the poor database.

On grounds of the variable moisture distribution in rocks, it makes little sense to define hydrological threshold values (like 91% or 50% saturation) from which a frost cycle is classified as effective or non-effective. Certainly, micro-gelivation is not likely to occur when the rock is far from saturation near below the surface. Even when the surface of the rock is dried out, there may be sufficient moisture in the deeper parts or in jointed areas, enabling frost-wedging to occur. However, it is necessary to bear in mind that higher conductivity along joints is probably not caused by a higher degree of saturation, but by a higher pore volume (and thus a higher water content) in fissured parts of the rock.

The spatially varying water-content leads to the assumption that water movement may play an important role in the frost weathering process. First measurements during subzero temperatures indicate that freezing instantly causes pronounced changes of the observed moisture distribution. In some cases water seems to be thrust away from near-surface ice lenses and cause increasing rock moisture in the surrounding area of the frozen parts. Further investigations are currently under way.

References

COUTARD, J.-P. & FRANCOU, B. (1989): Rock temperature measurements in two alpine environments: Implications for frost weathering. – Arct. Alp. Res. **21** (4): 399–416.

EVIN, M., FABRE, D. & JOHNSON, P. G. (1997): Electrical resistivity measurements on the rock glaciers of Grizzly Creek, St Elias Mountains, Yukon. – Permafrost Periglac. **8** (2): 181–191.

FAHEY, B. D. & LEFEBURE, T. H. (1988): The freeze-thaw weathering regime at a section of the Niagara Escarpment on the Bruce Peninsula, Southern Ontario, Canada. – Earth Surf. Process. Landforms **13**: 293–304.

HALL, K. (1986): Rock moisture content in the field and the laboratory and its relationship to mechanical weathering studies. – Earth Surf. Process. Landforms **11**: 131–142.

HALL, K. (1991): Rock moisture data from the Juneau Icefield (Alaska) and ist significance for mechanical weathering studies. – Permafrost Periglac. **2**: 321–330.

HALL, K. & HALL, A. (1996): Weathering by wetting and drying: Some experimental results. – Earth Surf. Process. Landforms **21**: 365–376.

HAUCK, C. (2001): Geophysical methods for detecting permafrost in high mountains. – VAW Mitt. 171, ETH Zürich.

HÖLLERMANN, P. (1964): Rezente Verwitterung, Abtragung und Formenschatz im oberen Suldental, Ortlergruppe. – Z. Geomorph. N. F., Suppl.-Bd. **4**.

LEHNERT, K. & ROTHE, K. (1962): Geophysikalische Bohrlochmessungen. – 300 p., Akademie-Verlag Berlin.

KING, L. (1982): Qualitative und quantitative Erfassung von Permafrost in Tarfala (Schwedisch-Lappland) und Jotunheimen (Norwegen) mit Hilfe geoelektrischer Sondierungen. – Z. Geomorph. N. F., Suppl.-Bd. **43**: 139–160.

KNEISEL, C. (2000): Anwendung der Gleichstromgeoelektrik zur Sondierung von Permafrost in jüngst eisfrei gewordenen Gletschervorfeldern des Oberengadins, Ostschweizer Alpen. – Jenaer Geogr. Schr. **9**: 39–50.

KNÖDEL, K., KRUMMEL, H. & LANGE, G. (1997): Geophysik. – Handbuch zur Erkundung des Untergrunds von Deponien und Altlasten, Vol. 3, 1063 p., Bundesanst. für Geowiss. und Rohstoffe, Springer-Verlag.

LOKE, M. H. (1999): Electrical imaging surveys for environmental and engineering studies – a practical guide to 2-D and 3-D surveys. – Copyright by M. H. Loke, Penang, Malaysia.

MATSUOKA, N. (1991): A model of the rate of frost shattering: Application to field data from Japan, Svalbard and Antarctica. – Permafrost Periglac. **2**: 271–281.

MCGREEVY, J. P. & WHALLEY, W. B. (1985): Rock moisture content and frost weathering under natural and experimental condition: A comparative discussion. – Arct. Alp. Res. **17** (3): 337–346.

MCGREEVY, J. P. & WHALLEY, W. B. (1987): Progress report: Weathering. – Progr. Phys. Geogr. **11** (3): 357–369.

SALVE, R. (1998): Near-surface wetting of a ponded basalt surface: observations using time domain reflectrometry. – J. Hydrol. **208**: 249–261.

SASS, O. (1998): Die Steuerung von Steinschlagmenge und -verteilung durch Mikroklima, Gesteinsfeuchte und Gesteinseigenschaften im westlichen Karwendelgebirge (Bayerische Alpen). – Münchener Geogr. Abh. **B 29**, 175 p.

TELFORD, W. M., GELDART, L. P. & SHERIFF, R. E. (1990): Applied Geophysics. – Cambridge University Press, 2. ed., 770 p.

THORN, C. E. (1979): Bedrock freeze-thaw weathering regime in an alpine environment, Colorado Front Range. – Earth Surf. Process. Landforms **4**: 211–228.

VORNDRAN, E. (1969): Untersuchungen über Schuttentstehung und Ablagerungsformen in der Hochregion der Silvretta (Ostalpen). – Schr. Geogr. Inst. Univ. Kiel **29** (3), 140 S.

WAGNER, S. (1996): DC resistivity and seismic refraction soundings on rock glacier permafrost in northwestern Svalbard. – Norsk Geogr. Tidsskr. **50**: 25–36.

WALDER, J. S. & HALLET, B. (1986): The physical basis of frost weathering: Toward a more fundamental and unified perspective. – Arct. Alp. Res. **18** (1): 27–32.

WEAST, R. C., ASTLE, M. J. & BEYER, W. H. (1989): Handbook of Chemistry and Physics. – Library of Congress, 69. ed. CRC Press, Boca Raton, Florida.

WHITE, S. E. (1976): Is frost action only hydration shattering? A review. – Arct. Alp. Res. **8** (1): 1–6.

Address of the author: Oliver Sass, Institut für Geographie, Universitätsstr. 10, D-86135 Augsburg, Germany.

Determining sediment thickness of talus slopes and valley fill deposits using seismic refraction – a comparison of 2D interpretation tools

Thomas Hoffmann and Lothar Schrott, Bonn

with 7 figures and 3 tables

Summary. During the last decade, shallow seismic refraction has been increasingly applied as a tool in geomorphological research. However, the interpretation of the data is often limited to one single method. The purpose of this paper is to highlight the discrepancies in subsurface models using different interpretation tools and to suggest appropriate methods for 2D modelling of complex terrains in alpine environments. Based on two interpreted profiles four different inversion and iterative methods (the intercept-time, wavefront-inversion, network-raytracing and travel-time inversion tomography) were applied. The intercept-time method should only be used where the assumption of a homogeneous and almost horizontally layered subsurface structure is reasonable. In more complex terrain the calculated depths of different layers may differ significantly (< 50%) compared to those obtained with wavefront-inversion or network-raytracing. In the case of gradual p-wave velocity changes and/or refractors of complex form the application of sophisticated methods is highly recommended. Wavefront inversion or network-raytracing allow interpretation of changing subsurface structures. In addition, the refraction tomography modelling is presented to visualize subsurface conditions with smoother transition zones, which are indicated by curved traveltime branches. Depending on the geomorphological conditions of the surface and subsurface, interpretation of seismic data may require a combination of different interpretation tools.

Zusammenfassung. Während der letzten Dekade wurde bei geomorphologischen Untersuchungen zunehmend die flachgründige Refraktionsseismik angewandt. In den meisten Fällen erfolgt die Interpretation der seismischen Daten jedoch nur mit einer Auswertungsmethode. Der Zweck dieser Arbeit liegt in der Dokumentation möglicher Abweichungen des modellierten Untergrundmodels unter Verwendung verschiedener Interpretationswerkzeuge und in der Vorstellung geeigneter Methoden zur 2D-Modellierung von komplexen Strukturen, wie sie beispielsweise in alpinen Tälern angetroffen werden. An zwei seismischen Profilen, die auf einem Schuttkegel und Alluvionen geschlagen wurden, sind verschiedene Inversions- und iterative Interpretatiosmethoden (Intercept-Zeit-Methode, Wellenfrontenverfahren, Network-Raytracing, Refraktionstomographie) getestet worden. Die Intercept-Zeit-Methode sollte nur bei homogenen und nahezu horizontal verlaufenden Schichtflächen verwendet werden. Bei komplexeren Untergrundstrukturen können die kalkulierten Tiefen der Refraktoren um über 50% von denen über das Wellenfrontenverfahren oder das Network-Raytracing ermittelten Werten abweichen. Im Falle von graduellen Geschwindigkeitswechseln der Schichten und/oder heterogen Refraktorverläufen wird die Anwendung von differenzierten Methoden empfohlen. Das Wellenfrontenverfahren oder das Network-Raytracing erlauben auch Interpretationen bei heterogen verlaufenden Schichtwechseln. In Ergänzung wird die Refraktionstomographie vorgestellt, die sich insbesondere bei gekrümmten Laufzeiten eignet und Untergrundstrukturen mit fließenden Übergängen visualisiert. Die jeweilige geomorphologische Situation und die zu erwartenden Untergrundverhältnisse erfordern unter Umständen eine Kombination der oben genannten Interpretationswerkzeuge.

1 Introduction

The use of shallow seismic refraction in geomorphological studies dates back to the early 1970s. At that time one channel seismographs with a limited number of scource-receiver combinations were generally used and interpretations were limited to simple subsurface models with homogeneous and horizontally layered structures (BARSCH 1973, WEISE 1972). In the 1980s, permafrost investigations – especially the determination of active layer thicknesses – led many researchers to apply geophysical techniques such as sledge hammer seismic refraction or DC-resistivity soundings (e.g. BARSCH & KING 1989, HAEBERLI 1985, KING 1982, VAN TANTENHOVEN & DIKAU 1990). These geophysical surveys provided useful informations regarding permafrost occurrence, active layer thickness and permafrost properties. With respect to seismic refraction, the interpretation was still usually restricted to the intercept-time method. In many cases, complex or more heterogenous subsurface structures were not adequately modelled, because of less powerful computer facilities and time-consuming calculations.

In the 1990s, the use of multi-channel seismographs and new interpretation software lead to the application of sophisticated interpretation tools including the Hagedoorn plus minus method, the generalized reciprocal method (GRM), the wavefront inversion method (WFI), the network-raytraycing and the refraction tomography (KNÖDEL et al. 1997, PALMER 1981, REYNOLDS 1997, VONDER MÜHLL 1993). These methods allow the interpretation of structured refractors and layers with varying p-wave velocities.

Over the last decade a variety of sophisticated interpretations and visualisations of the subsurface has been applied in several geomorphic studies (HAUCK 2001, HECHT 2000, HOFFMANN & SCHROTT 2002, VONDER MÜHLL 1993). Besides permafrost prospecting, a wider spectrum of geological and geomorphological settings and complex subsurface structures have been investigated. Applications of seismic refraction were reported from karst and loess covered landforms, valley fill deposits (talus slopes, alluvial fans and plains, avalanche cones etc.), block fields and landslides (HECHT 2000, HOFFMANN & SCHROTT 2002, SCHROTT et al. 2003, TAVKHELIDSE et al. 2000).

Comparisons between different interpretation tools applied at the same geomorphological location are still rare (SCHROTT et al. 2000). Under homogenous and quasi-horizontally layered subsurface structures the assumptions of the intercept-time method are met. On the basis of a single data set from a sledge hammer seismic survey, KING et al. (1992) showed the variation of two possible 1D interpretations using the intercept-time method.

Two different interpretations of the velocity lines of the second and third layer led to an increase in the thickness of the active layer from 4.1 to about 8 m. In the latter case the underlying permafrost might be relict and this has important implications to the geomorphological interpretation. In difficult and more complex terrain, the results may vary significantly, sometimes exceeding 50% and as a consequence, wrong geomorphological conclusions can be drawn from simplified geophysical interpretations. In such a case, simple interpretations using the intercept-time method remain unsatisfactory and will not provide a reliable model. SCHROTT et al. (2000) compared refractor depths from a blockfield using intercept method and GRM. Their results show that the intercept-time method may lead to differing results than the more sophisticated GRM. This is particularly true if the surface of the refractors is uneven. Therefore, they suggest that modelling results obtained using the intercept-time method be treated carefully. This shows the necessity of complementary and highly sophisticated interpretation tools to get more reliable models. In alpine environments, large talus slopes, alluvial plains and fans are the dominant geomorphic forms, but only a few reliable data are available concerning the depth to bedrock and the amount of sediment stored in these landforms. Seismic refraction surveys can provide substantial information for calculating sediment volumes and to differentiate sediment layers. Besides the problems which are concerned to the acquisition of the traveltime data, a major problem is the choice of the appropriate interpretation method to process the raw data. This paper focusses on the latter problem. Based on two seismic refraction surveys carried out in an alpine environment, the different interpretation tools are compared. In particular we focus on the application of

(i) intercept-time,
(ii) wavefront inversion,
(iii) network-raytracing, and
(iv) seismic refraction tomography.

We also discuss the different results in terms of estimating p-wave velocities and thicknesses of sedimentary strata.

2 Geomorphological setting and location of the seismic surveys

The seismic surveys discussed here were carried out on a talus cone and an alluvial plain in an alpine valley in the Bavarian Alps (Reintal, Wettersteingebirge) (Fig. 1). The area is dominated by a massive Triassic limestone. In the valley bottom, large talus slopes (sheets and cones), debris cones and alluvial plains have developed during Post-glacial times (SCHROTT et al. 2002). These sediment deposits in the valley are generally characterised by an increasing consolidation with depths. Seismic refraction profile RS62 was shot on a northfacing talus cone which lies below a steep rockwall (Fig. 1). The source area of this talus cone is the adjacent 1800 m high rockwall and sediment deposits which have been transported through gully erosion to the lower parts of the talus cone. The mean slope of the talus is about 21°. The upper end of the profile (distance = 0 m in Fig. 2, Fig. 3, Fig. 4 and Fig. 5) is located near the gully, whereas the lower end (69 m) lies on the central part of the talus cone.

Refraction sounding RS40 is located on an almost flat alluvial plain known as 'Hintere Gumpe' (Fig. 1) which was created after a rockfall event. The rockfall deposits dammed the

Fig. 1. Location of the refraction soundings RS40 and RS62. The "Hintere Gumpe" (bright area) is an alluvial plain, which was accumulated after the damming effect of a rockfall (hummocky terrain to the left of the Hintere Gumpe). The inset shows the location of the study site within the European Alps.

Partnach creek and a large sedimentation area has been developed (SCHROTT et al. 2002). It is likely that the investigated landforms of talus cone and alluvial plain consist of a mixture of different types of sediment sources (alluvial and rockfall deposits, till).

3 Methods

The seismic refraction surveys were carried out using a 24-channel Bison Galileo seismic system. We applied geophone spacings of 3 and 4 m resulting in profile lengths of 69 and 92 m, respectively. The energy source was a 5 kg sledge hammer and signals were stacked 5 times to improve signal to noise ratio. To allow the modelling of complex subsurface structures, 10 to 15 recordings were made along each profile.

The interpretation of the refraction sounding data was performed using the software package ReflexW V2.5 (SANDMEIER 2002). It allows the inversion of the traveltimes to p-wave velocity models of the underground. The interpretation of the models is based on the p-wave velocity of the different layers and on the calculated depths of the refractors. To analyse and to compare the gathered seismic data of two profiles we used

Fig. 2. Seismograms and intercept model of the forward (a) and reverse shot (b) of the refraction sounding RS62.

Determining sediment thickness 75

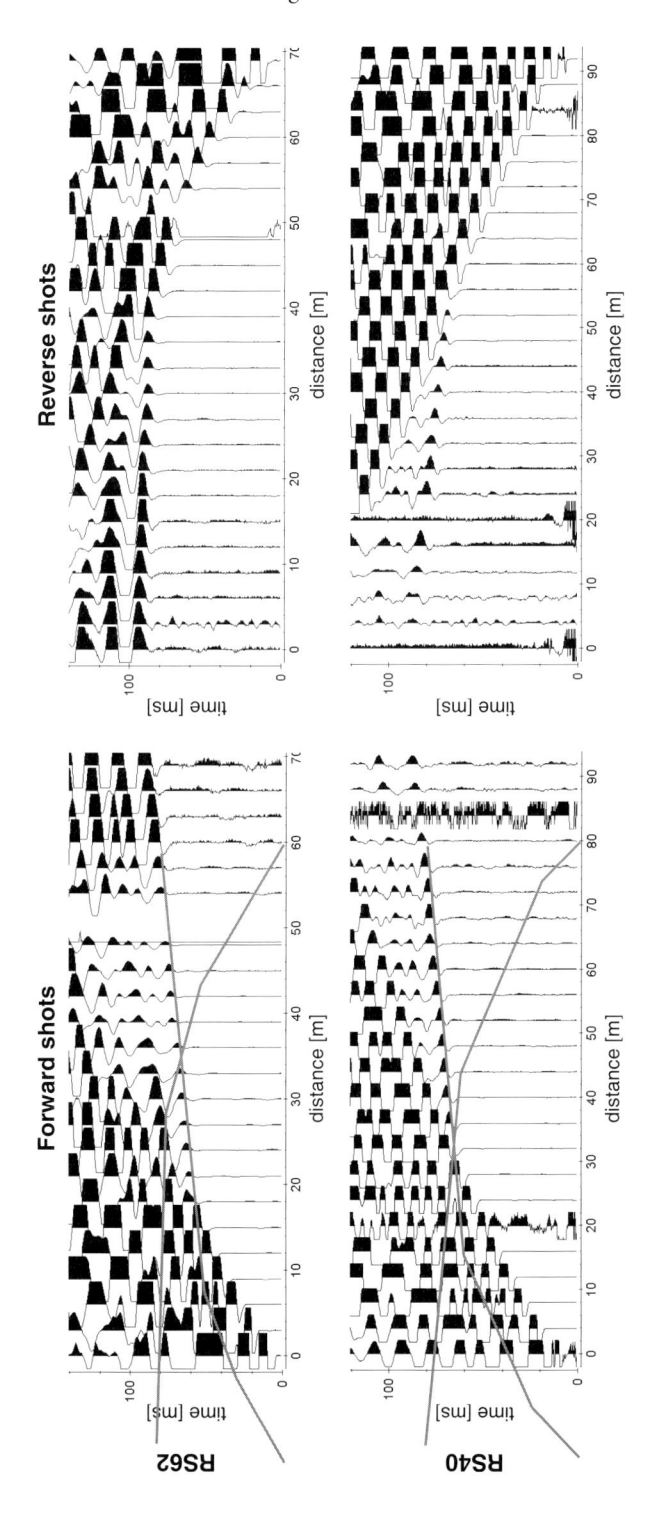

distance [m]

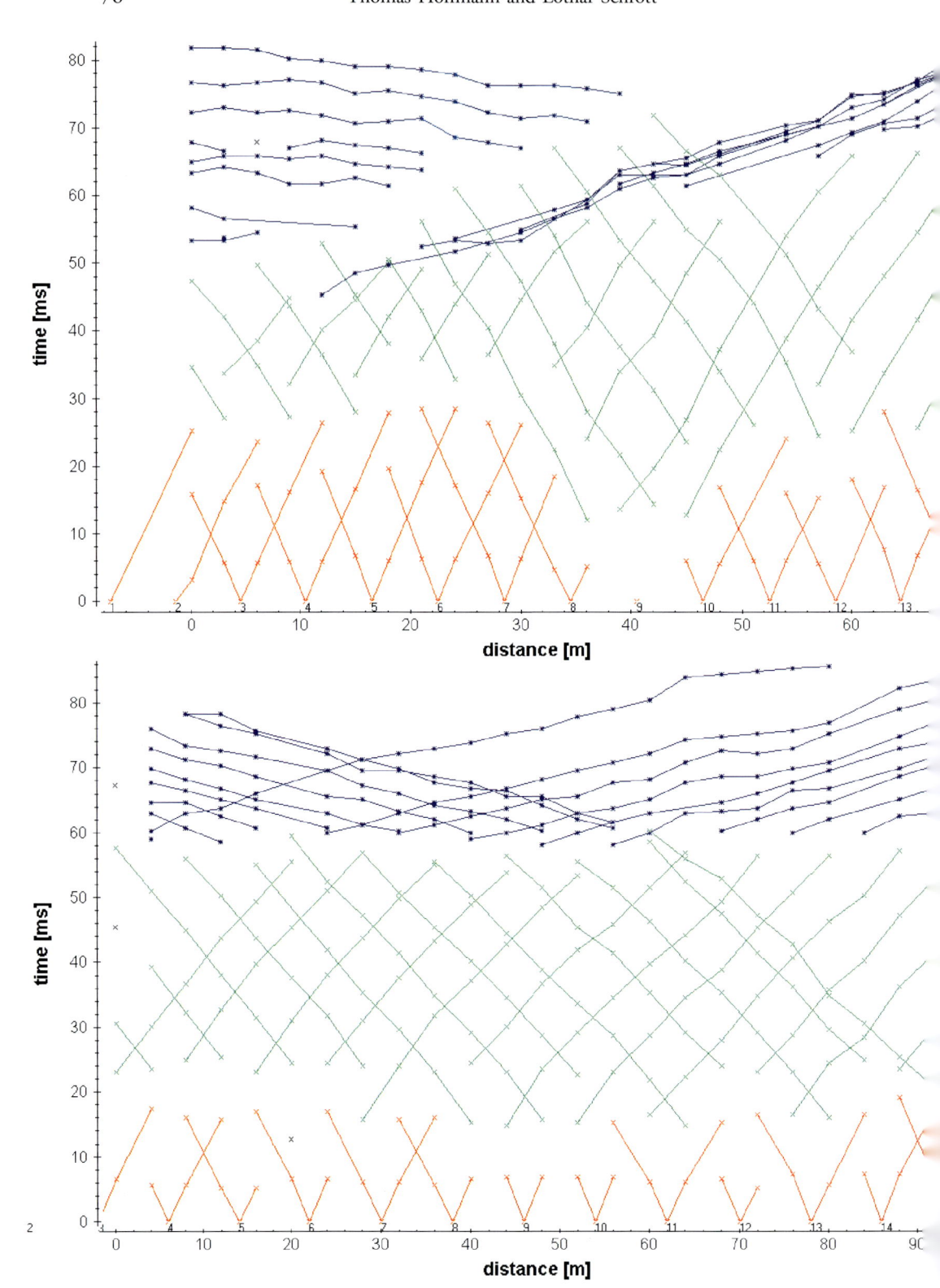

distance [m]

(i) the intercept-time method,
(ii) the wavefront-inversion method in combination with the network-raytracing, and
(iii) the seismic tomography method.

A detailed description of the interpretation tools is given by BRÜCKL et al. (1997) and SANDMEIER (2003).

The *intercept-time method* was only applied to forward and reverse shots, resulting in the apparent velocities and intercept-times of the different layers. The analysis of the intercept-time method is limited to one forward and one reverse shot in order to show the limitations of simple configurations and to highlight the differences to the subsequent listed methods. In this context it should be remembered that most seismic refraction data in early geophysical applications in geomomorphological research were analysed using the intercept-time method based on single forward and reverse shots. The layer thicknesses beneath the forward and reverse shotpoints and the 'real' velocities were calculated using the formulas described by BRÜCKL et al. (1997).

The *wavefront inversion modelling* was performed using the following steps (SANDMEIER 2003):
(i) Reading the travel-times of the first arrivals of the direct and refracted p-waves.
(ii) Combining travel-times of each record to a single travel-time diagramm (Fig. 3).
(iii) Assignment of the travel-times to the corresponding layers in the subsurface model (Fig. 3).
(iv) Inversion of the travel-times using the wavefront inversion algorithm.

In addition, WFI-models were evaluated by the network-raytracing method which allows the calculation of the travel-times of the WFI-models (SANDMEIER 2003, HECHT 2000). Therefore, WFI-models were manually modified until a better alignment of the calculated and the measured travel-times was obtained.

The *network-raytracing* was performed using the following steps:
(i) Choosing the start model (typically an existing WFI-model).
(ii) Loading the measured travel-times.
(iii) Calculation of the synthetic travel-times.
(iv) Visual comparison of the synthetic and the measured travel-times and manual modification of the subsurface model, primarily by changing the layer depths with the velocities of the start model kept constant.

Steps (iii) and (iv) were repeated until the differences between synthetic and measured travel-times were sufficiently small (< 4 ms).

The *tomography analysis* was applied for an automatic adaptation of synthetic travel-time data to real data. The tomography analysis allows the modelling of heterogenous subsurface structure with a minimum of knowledge about it. Therefore, we used tomography models as background information for the modification of the WFI-models to obtain a better correspondence of the synthetic and the measured travel-times. The algorithm is based on an iterative adaptation (SIRT-Simultaneous Iterative Reconstruction Technique). For a more detailed descriptions on the tomography parameter see Table 1.

Fig. 3. Assigned traveltimes of sounding RS62 (above) and sounding RS40 (below). The colors correspond to the assigned layers as followed: red = 1st layer, green = 2nd layer, blue = 3rd layer.

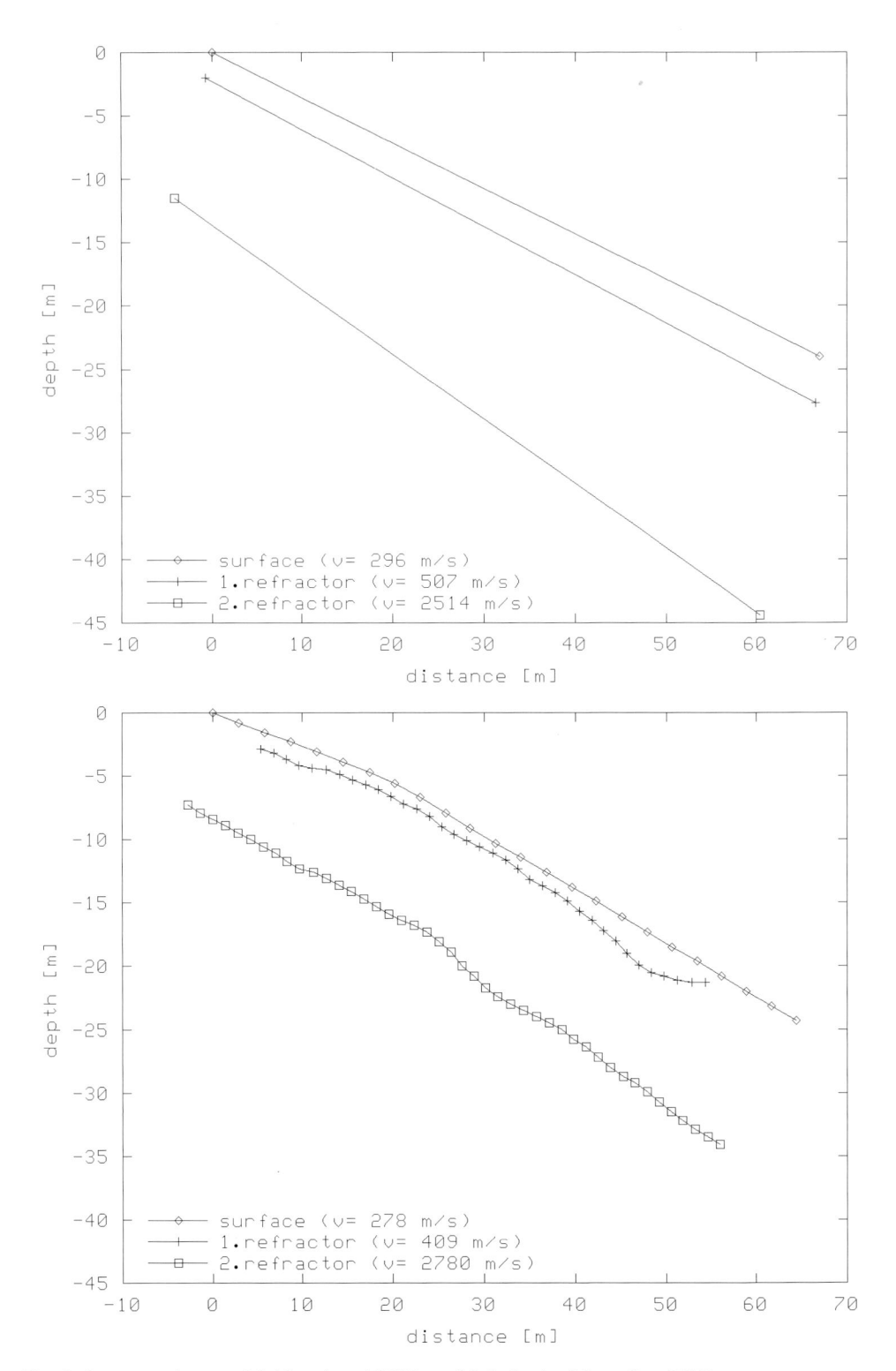

Fig. 4. Intercept-time model (above) and WFI model (below) of Sounding RS62.

4 Results

In a first step the incept-time method for the soundings RS62 and RS40 was applied. The apparent velocity, apparent depth and intercept-times of the survey were estimated by means of the first arrivals of forward and reverse shots (Table 2, Fig. 2). These data were used to calculate three-layer dipping models of the underground (Fig. 4 above and Fig. 7 above). The uppermost layer of the intercept-time model of the sounding RS62 is about 2.7 m thick with an increasing thickness down slope. The mean depth of the second refractor is 17 m with a dip of approximately 6.3° downslope relative to the surface. The velocities of the first two layers ($v_1 = 300$ m/s, $v_2 = 510$ m/s) are typical for unconsolidated debris with an increasing sediment compaction with depth. The velocity of the third layer ($v_3 = 2514$ m/s) indicates the base of the talus cone. The underlying material can be interpreted either as bedrock or highly consolidated till.

The corresponding models of the wavefront-inversion, the network-raytracing and the tomography analysis are shown in Fig. 4 and Fig. 5. The calculated layer velocities of the WFI and network-raytracing models ($v_1 = 280$ m/s, $v_2 = 410$ m/s, $v_3 = 2780$ m/s) are similar to the velocities given by the intercept-model. This results from the well defined (straight)

Table 1. Parameters used for tomography analysis.

Parameter	Value
Space increment	0.5 m
Stopping threshold (ratio of the actual and of the preceding decrease of the travel-time residual)	0.001
Maximum change of the model within one iteration	50 %
Start curved ray after iteration no.	4
Averaging model cells in x-direction (# of cells)	4
Averaging model cells in z-direction (# of cells)	0
Check no-ray area	Yes

Table 2. Results of intercept-time method for refraction soundings RS62 and RS40.

Sounding	Shot	Layer No.	Depth [m]	Velocity [m/s]	Intercept-time [ms]
RS62	forward	1	2.4	320	0
"	"	2	8.9	500	11.6
"	"	3	–	1810	40.4
"	reverse	1	2.9	270	0
"	"	2	15.1	520	18.2
"	"	3	–	4150	67.5
RS40	forward	1	2.3	360	0
"	"	2	14.7	640	10.4
"	"	3	–	3360	50.3
"	reverse	1	1.8	320	0
"	"	2	15.9	700	10.0
"	"	3	–	3100	49.9

Table 3. Comparison of mean depth, slope of layer base (refractor), and p-wave velocity of the corresponding layer using intercept-time and network-raytracing-models.

Sounding	Layer No.	Intercept			Network-Raytracing		
		mean depth [m]	mean slope [°]	velocity [m/s]	mean depth [m]	mean slope [°]	velocity [m/s]
RS62	1	2.7	− 1.2	300	1.6	0.1	280
"	2	16.7	− .6.3	510	10	− 4.9	430
"	3	–	–	2514	–	–	2710
RS40	1	2.1	≈ 0	340	1.7	− 0.4	340
"	2	20.4	≈ 0	670	14.7	− 0.5	710
"	3	–	–	3220	–	–	2740

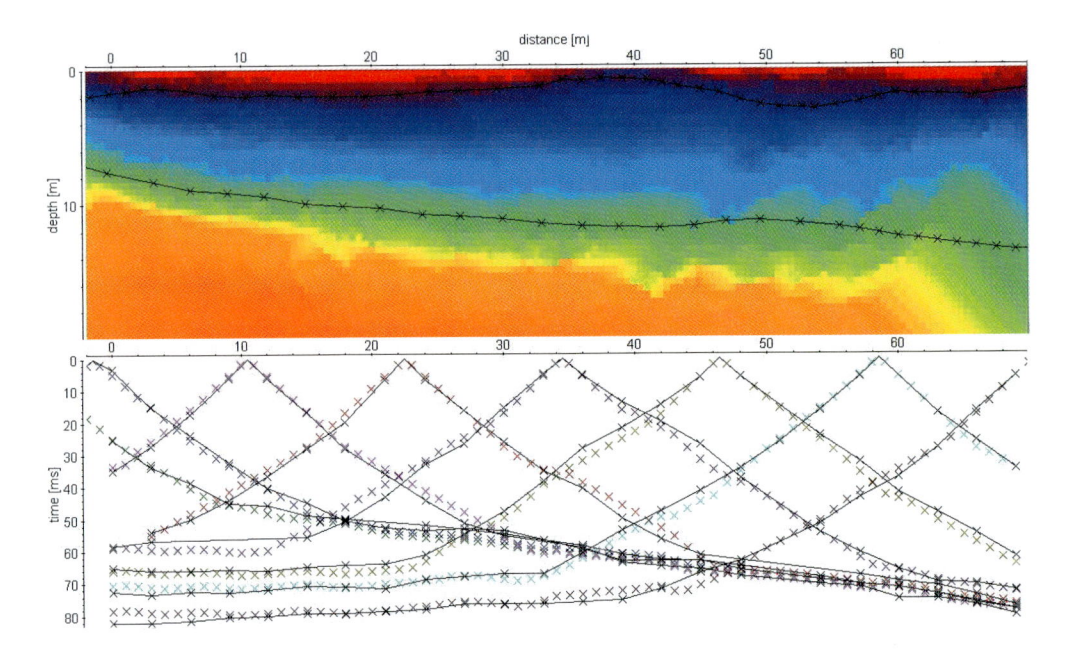

Fig. 5. Model results of network-raytracing and tomography analysis of the refraction sounding RS62 (talus cone); above: comparison of the network-raytracing model (black lines) and the refraction tomography (the colors indicate the velocities of the tomography model: red < 300 m/s, blue: 300–800 m/s, green and yellow: 800–1200 m/s, orange > 1200 m/s); below: comparison of the measured travel-times (black lines) and travel-times, which were calculated using the network-raytracing (colored crosses).

travel-time branches within the travel-time diagram (Fig. 3). Although the velocity of the third layer is at the lower limit of typical p-wave velocities known from limestone we interpret the second refractor as the bedrock base of the talus cone. This interpretation is supported by additional bedrock velocity measurements in the area. The high accuracy of the network-raytracing model is evidenced by the small differences (< 4 ms) between measured and calculated travel-times (Fig. 3 below). The subsurface underground structure obtained by the network-raytracing resembles the tomography model, as indicated by the almost identical course of the velocities of the refractors (Fig. 5 above). The first refractor of the network-raytracing model (upper black line in Fig. 5) for example marks the lower postion of the red part within the tomography model, which indicates velocites lower than 300 m/s. Furthermore, the slope of the second refractor (lower black line in Fig. 5) resembles the slope of the green belt which indicates a velocity between 800–1200 m/s.

The travel-time data of the sounding RS40 were interpreted as a three layer model. Using the intercept-time method the calculated velocities of the layers are 340 m/s (first layer), 670 m/s (second layer) and 3220 m/s (third layer) and corresponding refractor depths of approx. 2 and 20 m (Fig. 6 above). The refractor surfaces are parallel to the surface, i.e. with a slope that differs from the surface slope of less than 1°. A horizontal subsurface structure is also shown in the network-raytracing and the tomography model (Fig. 7 above). The mean depths of the refractors based on the the network-raytracing model are 1.7 m (first refractor) and 14.7 m (second refractor). The calculated velocities of the layers are 340 m/s (first layer), 710 m/s (second layer) and 2740 m/s (third layer).

5 Discussion

5.1 Interpretation of the subsurface models

Significant changes of sediment layers are not necessarily reflected by changes in p-wave velocities of refractors. Therefore, a genetic interpretation of different sediment layers requires additional information from drillings and sedimentological analysis. Based on our geomorphological and sedimentological analysis of sediment cores taken in the neighbourhood of the geophysical surveys the layers of the intercept-time-model as well as those of the network-raytracing model of RS 40 can be interpreted as followed:
- First layer: unconsolidated sedimentary material accumulated by fluvial processes.
- Second layer: mainly fine (clay, silt, sand) sediment of lacustrine and/or still water origin.
- Third layer: bedrock (limestone).

For a more detailed discussion on the sedimentological analysis see SCHROTT et al. (2002). The uppermost layer, with typical p-wave velocities smaller than 500 m/s, consists of unconsolidated debris (mainly pebbles) with a relatively small amount of sandy and silty sediment. The higher p-wave velocity (mainly between 500 and 1500 m/s) of the second layer is interpreted as a deposit with a high content of fine sediment (predominantly sand and silt) and higher compaction partly resulting from the overburden material. The estimated p-wave velocity of the lowest layer (generally > 2500 m/s) is characteristic for bedrock or highly consolidated glacigenic material.

Based on the p-wave velocities of the model layers we interpret the network-raytracing model of the sounding RS62 as follows:
- First layer: loose debris accumulated on top of the talus cone.
- Second layer: debris with a higher degree of compaction and a higher content of fines (compared to first layer).
- Third layer: bedrock (limestone).

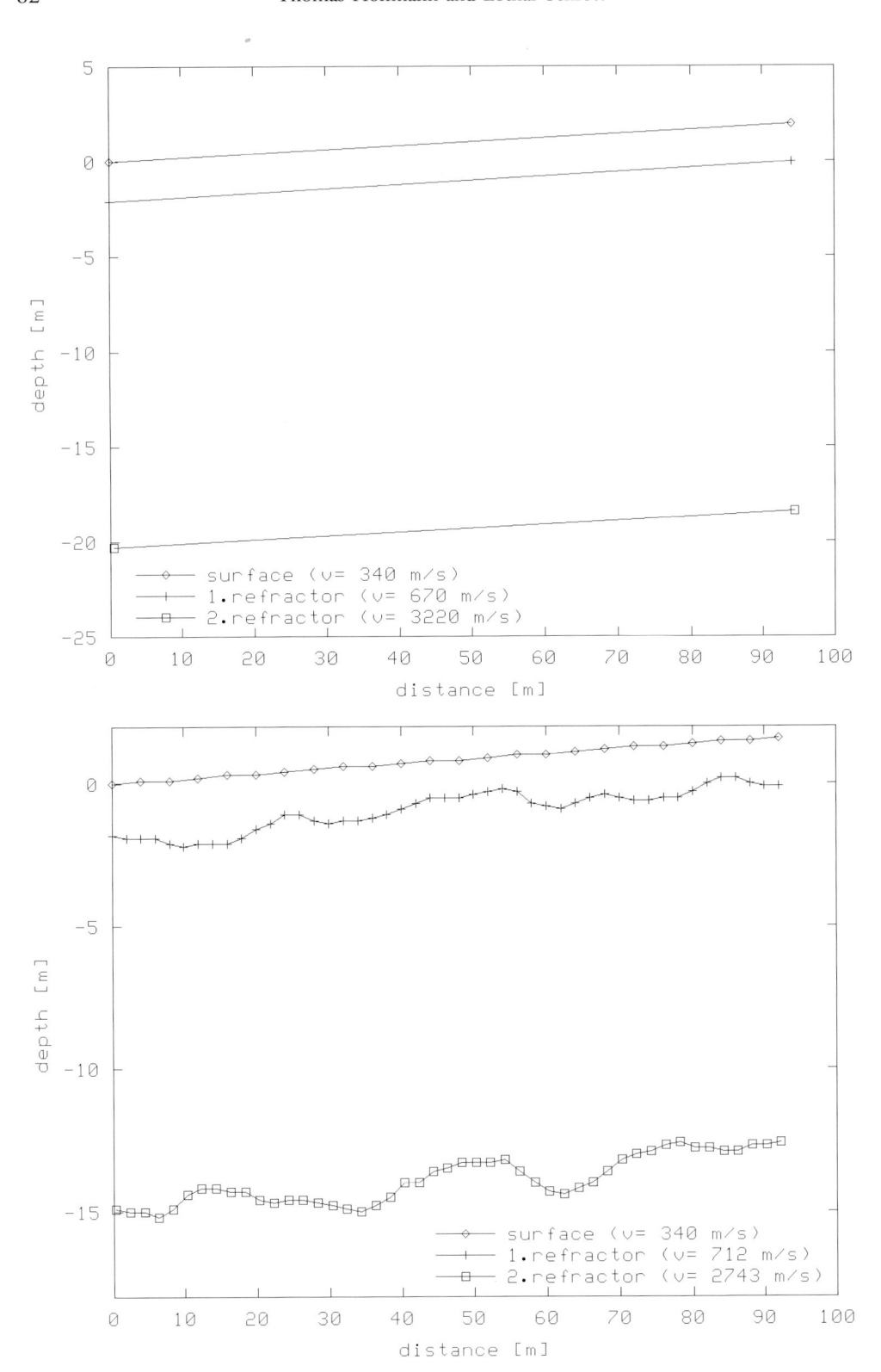

surface (v= 340 m/s)
1.refractor (v= 670 m/s)
2.refractor (v= 3220 m/s)

surface (v= 340 m/s)
1.refractor (v= 712 m/s)
2.refractor (v= 2743 m/s)

Based on the data discussed above, and on additional refraction soundings from the same study area, the three-layer model of the subsurface of alpine sediment storage types can be viewed in a more general sense. The uppermost layer with velocities smaller than 500 m/s and depths typically smaller than 2 m can be interpreted as the unconsolidated top layer with a high content of coarse debris. The second layer with p-wave velocities between 400 and 1500 m/s indicates a sedimentary stratum with a higher degree of compaction and a higher content of fine material. The third layer (generally faster than 2000 m/s) is characteristic for bedrock or glacigenic sediments: the differentiation of bedrock and glacial deposits using seismic refraction alone is not always possible (see also HOFFMANN & SCHROTT 2002). At this location, the influence of groundwater can be ruled out. Groundwater, however, can lead to both an increase in velocities from < 500 m/s to > 1200 m/s and so to an increase of noise in the raw data.

5.2 Comparison of the interpretation methods

Due to the simple geometric assumptions of the intercept-time method, reliable interpretations can only be achieved if the slope angle of the refractors relative to the surface is less than 10° (BRÜCKL et al. 1997). This is valid for all analysed models.

To compare the network-raytracing models with the intercept-time models, the mean depth and the slope of the network-raytracing model layers were calculated using a linear regression of each layer. The calculated layer depths of the intercept-time models are always larger than those of the network-raytracing models. We assume an error in depth calculation of ±20% (HOFFMANN 2002). Therefore, the differences between the depths of the layers derived from different procedures are statistically significant. Because more data are used in the network-modelling and because of the similarity between the measured and calculated travel-times, the depths of the network-raytracing models give a better spatial resolution than those of the intercept-time models. The velocity differences between the two models are less than 16%, and are therefore in a good agreement, whereas the velocities of the tomography models are much lower at the refractor depths (Fig. 5 and Fig. 7). This can be explained by the smaller velocity gradients of the tomography analysis, due to smoothing in the tomography algorithm and the deeper penetration of the refracted waves compared to the network-raytracing analysis (SANDMEIER, pers. commun.).

A similar course of the refractors of the interpreted underground models generated by the network-raytracing and the tomography can be observed (Fig. 5 and Fig. 7). The distribution of red color (corresponding to velocities of the tomography below 400 m/s) is similar to the course of the first refractor (Fig. 5). The second refractor in Fig. 5 shows the same slope as the light green color band within the tomography model. We therefore suggest that the tomography model is a helpful background information for modifying the WFI models, and to obtain subsurface models with a smaller difference between observed and calculated travel-times. The differences between the network-raytracing model and the tomography between 55 and 70 m are effects of the border of the tomography model.

Fig. 6. Intercept-time model (above) and WFI model (below) of Sounding RS40.

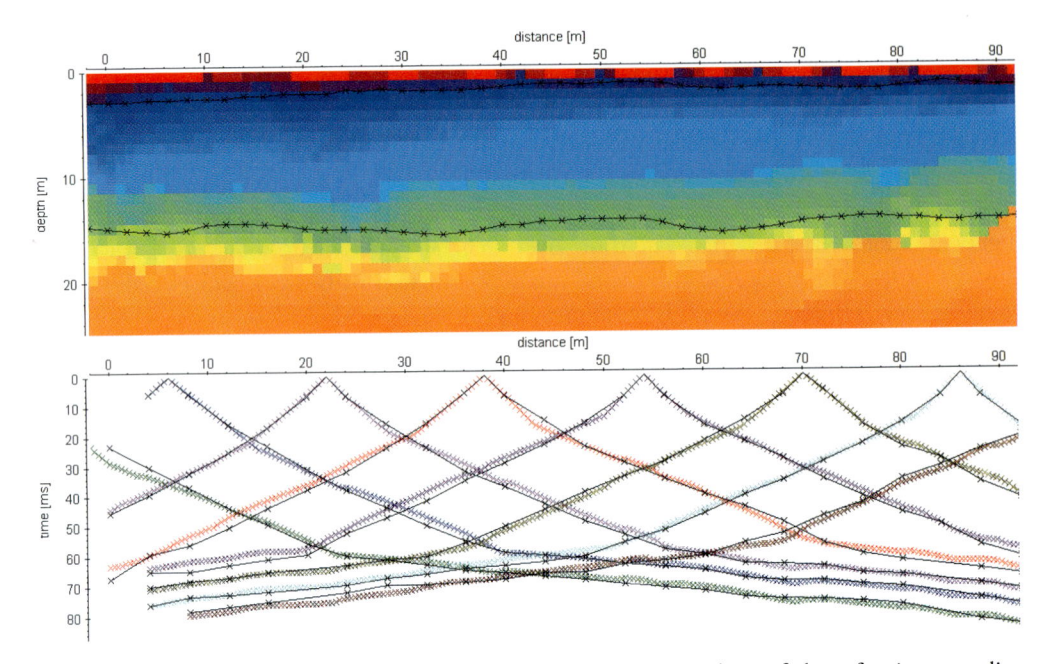

Fig. 7. Model results of the network-raytracing and tomography analysis of the refraction sounding RS40 ('Hintere Gumpe'); above: comparison of the network-raytracing model and the refraction tomography (the colors indicate the velocities of the tomography model: red < 300 m/s, blue: 300–800 m/s, green and yellow: 800–1200 m/s, orange > 1200 m/s; and below: comparison of the measured travel-times (black lines) and calculated travel-times (colored crosses).

6 Conclusions

Depending on the geomorphological condition of the surface and subsurface the interpretation of seismic data may require a combination of different tools. Here we used seismic refraction data of a talus cone and an alluvial plain in a mountain environment. Our modelling results which are based on the intercept-time method, network-raytracing, wavefront-inversion, and tomography showed the following advantages and disadvantages:

- The *intercept-time method* as described above is a useful tool to obtain a simple and quick picture of the underground. The limitations of the method are set by the assumption of a subsurface structure with homogenous layers and simple refractor surfaces without getting a detailed resolution, and by the frequently use of a limited number of data. Our results show that the calculated depths may differ significantly from those calculated by more complex methods. The application of this method is therefore only recommended for simple approximations based on limited information (e.g. only one forward and reverse shot) or for sites with a relatively simple subsurface structure.
- *Wavefront-inversion* was used as a strong tool to create a starting model for the network-raytracing. Major limitations of this method are connected to the assignment of the travel times to specific layers. For example, in the case of a gradual increase of the p-wave velocity with depth a definite assignment is not possible.

- *Network-raytracing* allows the modelling of complex subsurface structures by comparing measured and calculated travel-times. This offers also the possibility of a quality check of the underground model. All existing data may be used in the model interpretation. However, the interpretation of complex travel-times and the adaptation of the models needs in many cases expert knowledge and is very time consuming. In order to minimize these limitations, the starting model for the network-raytracing procedure was based on a wavefront-inversion model. The wavefront-inversion model was then slightly modified to adjust the measured and calculated travel-times. The network-raytracing analysis may lead to non-unique solutions, depending strongly on the knowledge and experience of the person doing the analysis.
- The *tomography method*, like network-raytracing, allows the modelling of complex structures of the underground. Since no assignment of travel-times to layers is necessary, it is relatively easy to handle even if the interpreter has little modelling experience (SANDMEIER 2003). The resulting subsurface interpretation is always a smoothed non-layered, gradient model which is not suitable for accurately estimating layer thicknesses. However, it gives a good picture of changing underground structures. The tomography model helped to recognise more easily the subsurface structures and is therefore a useful tool to guide the modification of the subsurface models for the network-raytracing analysis. The data show a good agreement of the subsurface model structure using both, network-raytracing and tomography models.

A combination of the wavefront-inversion, network-raytracing and tomography methods is producing a more complete picture of subsurface conditions.

To summarise, the following interpretation strategy was applied:
- Intercept-time method with single forward and reverse shot: quick interpretation of travel-times.
- Wavefront-inversion: interpretation of the traveltimes for non linear (structured) refractors.
- Network-raytracing: validation and eventually modification of wavefront-inversion models.
- Tomography method: as a tool for recognizing the underground structure. In the case of missing boundaries the refraction tomography is the only applicable method of those described here.

The results showed clearly the limitations and possible errors of the single application of the intercept-time method, which has normally been used in previous geomorphological studies. To transfer raw data into high quality subsurface data and to minimize model errors, the combined use of these tools in seismic refractions surveys is highly recommended. Existing subsurface data which are based on only single shots and interpretation by the intercept-time method must be treated with care.

Acknowledgements

We thank Gabi Hufschmidt, Martin Hankammer, André Niederheide and Sandra Amlang for their assistance in the field. Charly Wehrle and Karl Woerndle provided many logistical support to transport the equipment into the alpine valley. The corrections and comments of Scott Stewart, Nel Caine and the two referees Magnus Friberg and Horst Rüter are greatly appreciated. Financial support was provided by a grant from the Deutsche Forschungsgemeinschaft (DFG).

References

Barsch, D. (1973): Refraktionsseismische Bestimmung der Obergrenze des gefrorenen Schuttkörpers in verschiedenen Blockgletschern Graubündens, Schweizer Alpen. – Z. Gletscherkde. Glaziol. **9** (1–2): 143–167.

Barsch, D. & King, L. (1989): Origin and geoelectrical resistivity of rock glaciers in semi-arid subtropical mountains (Andes of Mendoza, Argentinia). – Z. Geomorph. N. F. **33**: 151–163.

Brückl, E., Kirchenheimer, F., Krummel, H., Liebhart, G., Reimers, L., Sandmeier, K. J. & Utecht, T. (1997): Refraktionsseismik. – In: Knödel, K., Krummel, H. & Lange, G. (eds.) (1997): Geophysik. Handbuch zur Erkundung des Untergrundes von Deponien und Altlasten. – pp. 510–599, Springer.

Haeberli, W. (1985): Creep of mountain permafrost: internal structure and flow of alpine rock glaciers. – Mitt. VAW, ETH-Zürich, 143 pp.

Hauck, C. (2001): Geophysical methods for detecting permafrost in high mountains. – Mitt. VAW, ETH-Zürich, 150 pp.

Hecht, S. (2000): Fallbeispiele zur Anwendung refraktionsseismischer Methoden bei der Erkundung des oberflächennahen Untergrundes. – Z. Geomorph. N. F., Suppl.-Bd. **123**: 111–123.

Hecht, S. (2003): Differentiation of loose sediments with seismic refraction methods – potentials and limitations derived from case studies. – Z. Geomorph. N. F., Suppl.-Bd. **132**: 89–102.

Hoffmann, T. (2002): Modellierung von Sedimentmächtigkeiten und Hangrückverwitterungsraten in einem alpinen Einzugsgebiet unter Anwendung der Refraktionsseismik (Reintal, Bayerische Alpen). – Diploma thesis, Department of Geography, University of Bonn, 85 pp.

Hoffmann, T. & Schrott, L. (2002): Modelling sediment thickness and rockwall retreat in an Alpine valley using 2D-seismic refraction (Reintal, Bavarian Alps). – Z. Geomorph. N. F., Suppl.-Bd. **127**: 175–196.

King, L. (1982): Qualitative und quantitative Erfassungen von Permafrost in Tarfala und Jotunheimen mit Hilfe geoelektrischer Sondierungen. – Z. Geomorph. N. F., Suppl.-Bd. **43**: 139–160.

King, L., Gorbunov, A. P. & Evin, M. (1992): Prospecting and mapping of mountain permafrost and associated phenomena. – Permafrost Periglac. Process. **3** (2): 73–81.

Knödel, K., Krummel, H. & Lange, G. (1997): Geophysik. Handbuch zur Erkundung des Untergrundes von Deponien und Altlasten. – Springer-Verlag, Berlin, 1063 pp.

Palmer, D. (1981): An introduction to the generalized reciprocal method of seismic refraction interpretation. – Geophysics **46** (11): 1508–1518.

Reynolds, J. M. (1997): An introduction to applied and environmental geophysics. – John Wiley & Sons, Chichester, 778 pp.

Sandmeier, K. J. (2002): Reflexw V2.5. Manual. – Sandmeier Scientific Software, Karlsruhe, 264 pp.

Sandmeier, K. J. (2003): Refraction seismics demo, Reflex W, version 3.0 – Z. Geomorph. N. F., Suppl.-Bd. **132**: CD-ROM.

Schrott, L., Niederheide, A., Hankammer, M., Hufschmidt, G. & Dikau, R. (2002): Sediment storage in a mountain catchment: geomorphic coupling and temporal variability (Reintal, Bavarian Alps, Germany). – Z. Geomorph., Suppl.-Bd. **127**: 175–196.

Schrott, L., Hufschmidt, G., Hankammer, M., Hoffmann, T. & Dikau, R. (2003): Spatial distribution of sediment storage types and quantification of valley fill deposits in an alpine basin, Reintal, Bavarian Alps, Germany. – Geomorphology **55** (1–4): 1–19.

Schrott, L., Pfeffer, G. & Möseler, B. M. (2000): Geophysikalische Untersuchungen an einer Blockhalde im Mittelgebirge (Hundsbachtal, Eifel). – Acta Universitatis Purkynianae 52, studia biologica **4**: 19–31.

Tavkhelidse, T., Schulte, A., Stumböck, M. & Schuhkraft, G. (2000): Aufbau und Entwicklung der Schuttkegel im Finkenbachtal, Südlicher Odenwald. – Jenaer Geogr. Schr. **9**: 95–110.

Van Tantenhoven, F. & Dikau, R. (1990): Past and present permafrost distribution in the Turtmanntal, Wallis, Swiss Alps. – Arct. Alpine Res. **22** (3): 302–316.

Vonder Mühll, D. (1993): Geophysikalische Untersuchungen im Permafrost des Oberengadins. – VAW-Mitteilungen, ETH-Zürich, 183 pp.

Weise, O. R. (1972): Zur Bestimmung der Schuttmächtigkeit auf Fußflächen durch Refraktionsseismik. – Z. Geomorph. N. F., Suppl.-Bd. **14**: 54–65

Address of the authors: Dipl. Geograph Thomas Hoffmann and Dr. Lothar Schrott, Universität Bonn, Geographisches Institut, Meckenheimer Allee 166, D-53115 Bonn, Germany.

Differentiation of loose sediments with seismic refraction methods – potentials and limitations derived from case studies

Stefan Hecht, Heidelberg

with 11 figures

Summary.   Results from various field studies demonstrate the potentials and limitations of seismic refraction methods for the investigation of the shallow subsurface focussing on the differentiation of loose sediments. Vertical differentiation of loose sediments is often made difficult by the occurrence of 'hidden layers' which can be illustrated by the results of the case study 'Bietigheim'. Nevertheless, any layer above a 'hidden layer' can be detected and the supplementary application of drillings or other geophysical investigations can help to improve the results. In contrast, the detection of lateral velocity contrasts is a great advantage of seismic refraction 2D-methods. The results of the case study 'Kirchener Tal' show lateral changes of slope deposits as well as the palaeosurface of the underlying limestone. The delimitation of loose sediments from bedrock normally profits from large contrasts in seismic velocities. Thus 2D-measurements provide the reconstruction of topography in bedrock. Furthermore, the data from the case study 'Maustobel' illustrate the possibilities concerning the delimitation of slide masses from bedrock. Another example shows a 3D-display of the layer configuration including the delimitation of loose sediments to bedrock and the differentiation of the loose sediments themselves. Thus seismic refraction methods provide useful information about landscape evolution, making them valuable tools for geomorphological investigations.

Zusammenfassung.   Möglichkeiten und Grenzen bei der Anwendung refraktionsseismischer Methoden – Fallstudien zur Untersuchung von Lockersedimenten. Anhand mehrerer Fallbeispiele werden Grenzen und Möglichkeiten der Anwendung refraktionsseismischer Methoden zur Erkundung des oberflächennahen Untergrundes aufgezeigt. Dabei steht die Abgrenzung von Lockersedimenten im Vordergrund der Untersuchungen. Die vertikale Differenzierung von Lockersubstraten wird zwar – wie im Fallbeispiel 'Bietigheim' – häufig von dem Auftreten sog. 'hidden layers' erschwert, trotzdem können zumindest alle Schichtgrenzen über einem 'hidden layer' bestimmt werden. Zusätzliche Bohrungen oder Ergebnisse anderer geophysikalischer Methoden können die Auswertungen entscheidend verbessern. Dagegen bietet die Unterscheidung lateraler Geschwindigkeitsänderungen das größte Potenzial für refraktionsseismische 2D-Methoden. Das zeigen v.a. die Ergebnisse des Fallbeispiels 'Kirchener Tal'. Dort konnten sowohl laterale Änderungen der quartären Lockersedimente nachgewiesen werden wie auch der Verlauf eines heute überdeckten pleistozänen Donau-Prallhanges.

Da die Abgrenzung von Lockersedimenten zum Festgesteinsuntergrund i.d.R. von großen Geschwindigkeitsunterschieden der seismischen Wellenausbreitung begünstigt wird, ist es mit Hilfe refraktionsseismischer 2D-Methoden möglich, Topographien im Festgestein relativ genau nachzuzeichnen. Darüber hinaus verdeutlicht das Fallbeispiel 'Maustobel' die Möglichkeiten bei der Abgrenzung von Rutschkörpern vom darunter liegenden Festgestein. Insgesamt stellen die Ergebnisse refraktionsseismischer Messungen wichtige Informationen über die Landschaftsentwicklung zur Verfügung und werden damit zu einem unverzichtbaren Instrument geomorphologischer Untersuchungen.

1 Introduction

In this paper results from various seismic field studies are presented, focussing on the vertical and horizontal differentiation of loose sediments and the delimitation to the underlying bedrock. Methodological problems as well as the potentials and advantages of using seismic refraction methods for the purpose of geomorphological investigations are discussed.

The measurements took place with a 24-channel seismograph (BISON) and a sampling rate of 0.2 ms. Geophone intervals of 2 to 5 m were selected using a total of 24 traces. Each geophone spread had several offset shots at each end of the spread and several inline shots, at least midway and at every quarter of each spread. If more than one spread was used, each spread overlapped the next one by several geophone positions. A shorter geophone spacing and additional shots per spread helped to resolve complex layer configurations. A sledge hammer was used to generate the P-waves.

Data was processed with the software package REFRA by SANDMEIER & LIEBHARDT and the modules for refraction seismics of the software package REFLEX by SANDMEIER (SANDMEIER 2000). A combination of different interpretation tools – such as interactive 1D-modelling, the Generalized Reciprocal Method (GRM) developed by PALMER (1981) and network raytracing – was used for the reconstruction and modelling of the shallow subsurface. The GRM was used to generate an initial model of the subsurface. This method is discussed controversially in literature. According to SHARMA (1997) the GRM is particularly suitable for situations with lateral changes in layering. In contrast, WHITELY (1992) and SJÖGREN (2000) consider the results of the GRM to be partly incorrect concerning depth and velocity information.

For this reason, network raytracing (SANDMEIER 2000) was applied to verify and improve the calculated 2D-depth models extracted by the combination of 1D-modelling with the GRM. In the first step every model must be gridded with an appropriate spatial increment. Then the raypath with the shortest traveltime from one point to the next is determined by the appliance of the Dijkstra algorithm (SANDMEIER 2000). The comparison of the resultant computed travel times with the measured travel times allows a reliable visual control to estimate the quality of the appropriate model. Because of this advantage, great emphasis has been laid on the application of this method. In the following examples from field studies, computed ray paths of seismic wave propagation are used to illuminate some of the advantages and some of the problems of seismic refraction modelling.

2 Differentiation of loose sediments

As a basic principle, the seismic velocity of P-waves (v_p) of almost every media varies over a huge range, making it impossible to determine a specific layer only by its seismic velocity (Fig. 1). In limestone, for example, or dolomite, velocities may occur from 2000 to more than 6000 m/s, depending on jointing or the degree of weathering. Specific characteristics of rock physics are presented by MAVKO et al. (1998). For some materials (e.g. siltstone), not enough data are documented to indicate the whole range of velocities which can be found in consistent geological strata.

Investigations of the shallow subsurface with seismic refraction methods often deal with loose sediments, which are in many cases very heterogeneous. For example, the grain size distribution or the level of consolidation of quaternary sediments may vary on a small scale. Therefore, the resultant seismic velocities of these layers may also differ on a small scale. This influences the interpretation of seismic traveltime data, because errors made in the determination of the uppermost layer directly affect the modelling of any layer below (SANDMEIER 1997). On the other hand side, changes in the velocity of the uppermost layer(s) provide useful information about essential lateral variations of the layer configuration itself. For detailed information on basic principles of seismic refraction methods refer to

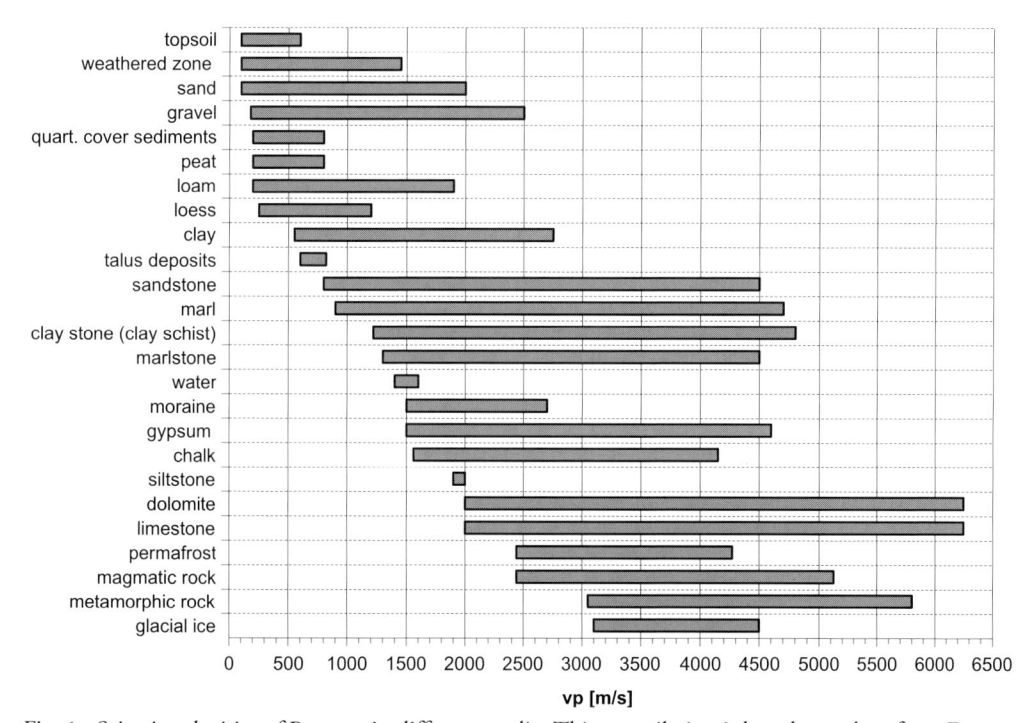

Fig. 1. Seismic velocities of P-waves in different media. This compilation is based on values from FERTIG (1997) and was amended with additional data from numerous authors. Indicated are the ranges in which data was measured. In some cases small ranges are shown because of a lack of sufficient data (e.g. siltstone).

REYNOLDS (1997), SHARMA (1997), SANDMEIER (1997), SANDMEIER & LIEBHARDT (1997), KIRSCH & RABBEL (1997) and BURGER (1992). Applied considerations about these methods are documented in HECHT (2000, 2001), EBERLE et al. (2001) and KRAMMER (1997), for example.

Lateral changes of the seismic velocities are not the only factor complicating the correct interpretation of the data. Sometimes the vertical configuration of layering can not be resolved by seismic refraction methods. In most cases the seismic velocity increases with increasing depth. In contrast, 'low velocity layers' are the result of a velocity decrease with increasing depth. For these layers no first breaks can be determined in the corresponding seismograms. Thus they are not detected with standard interpretation tools. Furthermore layers appear as so called 'hidden layers' and no traveltime information can be measured from these layers either. This phenomenon is due to a weak contrast of the seismic velocities of different layers or to a low thickness. As a consequence, the depth interpretation of the layers beneath is incorrect. Nevertheless, LANKSTON (1990, 1989) pointed out, that even if 'hidden layers' exist, at least all layers above them can be determined exactly. Moreover, additional data from drilling results or from other geophysical investigations can be taken into account to improve the fragmentary results of such seismic measurements.

Another problem may occur if the seismic velocity of the uppermost layer is lower than the airborne sound (330 m/s). In that case the designation of first breaks is problematic and may result in inaccurate interpretations of any underlying layer (Fig. 2). SCHANZ (1996) also reported such difficulties concerning a concrete field study.

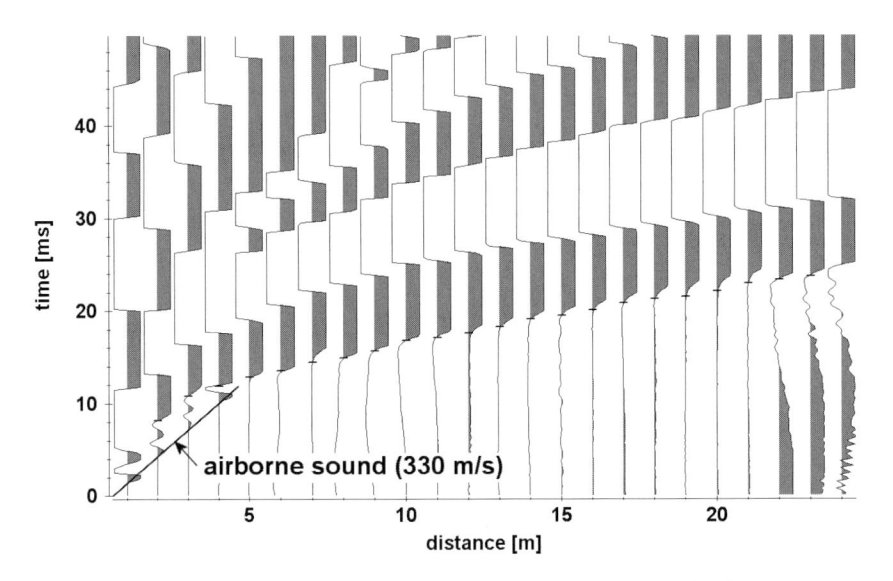

Fig. 2. Sketch of a seismic section. The first breaks of seismic waves close to the shotpoint at 0,5 m are due to the airborne sound. Consequently, the velocity of P-waves in the uppermost layer must be slower. In such a case it is not possible to determine the seismic velocities in these layers exactly and therefore the interpretation of the underlying layers is incorrect.

2.1 Vertical differentiation of loose sediments

Problems concerning the vertical differentiation of loose sediments with seismic refraction methods are illustated by the results of two case studies that are located in the southwest of Germany. First an example is presented that focuses on the 'hidden layer' problem (case study 'Bietigheim') and the following case study ('Untergrombach') is linked up to the occurrence of groundwater in alluvial sediments.

2.1.1 Case study 'Bietigheim': The 'hidden layer' problem

Field studies on a loess-covered pleistocene terrace of the River 'Enz' near Bietigheim (Baden-Württemberg, Germany) illustrate the difficulty with 'hidden layers' (Fig. 4). The principle task was the differentiation of loess and fluvial sediments (sand and gravel) that appear in differing thicknesses over solid limestone. The fluvial sediments are therefore only partly detectable along the seismic profile.

The measurements took place with a geophone spacing of 2 m with several offset and inline shots. Fig. 3 shows the travel time curves of selected shots. The travel time sections along the profile clearly point at lateral changes in the layering which could be approved by the computed raypaths of refracted wave propagation through the final layer model (Fig. 4). In Figs. 4a and 4b refracted rays occur at every interface. Thus all layers (topsoil, loess, sand/gravel, limestone) can be recognised by the measurements. Below the shotpoint in Fig. 4c, the layer of sand/gravel is rather thin and the limestone is relatively near the surface. As a result the seismic waves refracted at the limestone outrun the waves refracted at the fluvial sediments. Thus no traveltime data can be recorded from this interface. The layer of sand and gravel is characterised as a 'semi-hidden layer' according to SJÖGREN (1984). In this case the reconstruction of the layer configuration is only possible by the consideration of numerous shots along and outside the seismic profile. Otherwise data interpretation leads to ambiguous results.

2.1.2 Case study 'Untergrombach': The occurrence of ground-water in alluvial sediments

Fig. 5 shows another problem concerning seismic refraction investigations in loose sediments. The main objective was the detection of the ground-water table and the vertical differentiation of the alluvial fill. But the occurrence of the groundwater-table in the loose sediments conceals the existence of any layer below of slower seismic velocity than the P-waves in ground-water. Thus it is not possible to differentiate the water-saturated alluvial fill. But clear changes in the seismic velocities enables the differentiation of topsoil/colluvium (lateral variations of v_p range between 160 and 220 m/s), alluvial fill (730 m/s) and the ground-water table (1770 m/s). The depth of the water-table matches the water level data from wells close to the seismic profiles almost exactly. The measurement was made up of four overlapping seismic profiles with geophone spacings of 1 and 3 m. Each spread had several offset and inline shots.

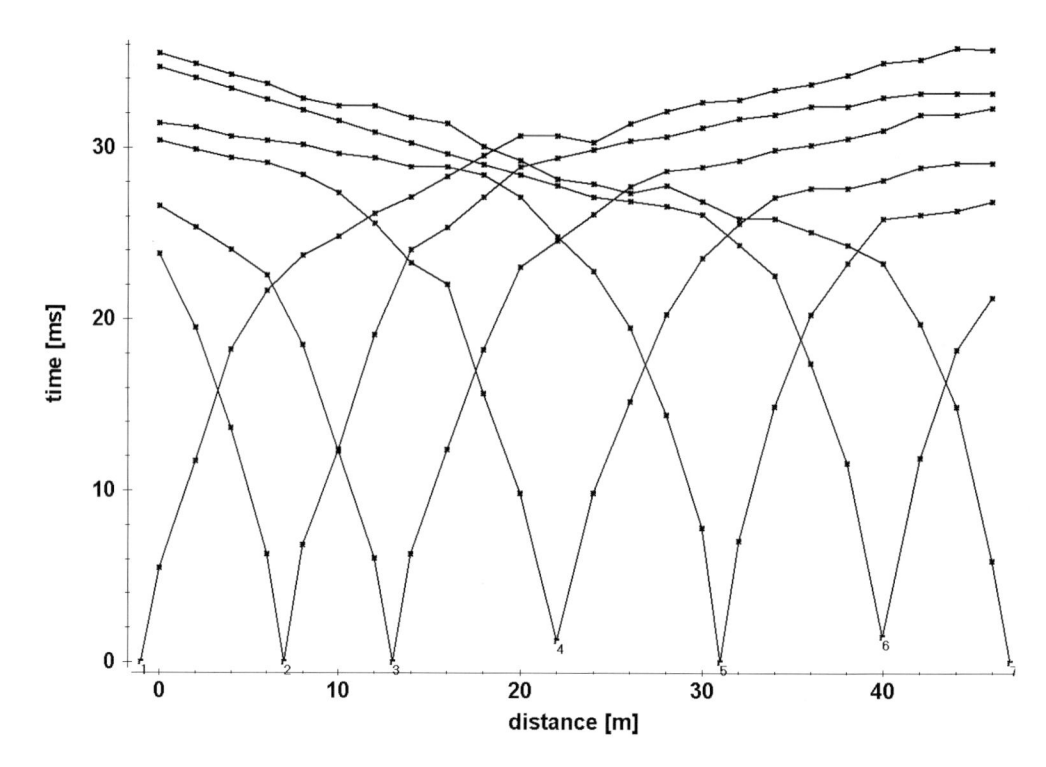

Fig. 3. Travel time curves of selected shots along profile "Bietigheim". Note the different slopes of the traveltime sections between 20 and 30 ms that indicate the lateral changes in the layering displayed in Fig. 4.

2.2 Lateral differentiation of loose sediments

'Hidden layers' as described above, often lead to inaccurate interpretation concerning the depths of the layer interfaces. Thus vertical differentiation of strata is complicated in many cases. In contrast, lateral changes in layering can be detected with high precision, which is a great advantage of seismic refraction methods.

2.2.1 Case study 'Kirchener Tal': Investigations in a former valley of the River Danube

Fig. 6 shows a complex layer configuration at a former valley of the River Danube near Ehingen (Baden-Württemberg, Germany). Limestone at the basement is partly covered by sand/gravel, a weathered zone/redeposition zone and marlstone. The different strata partially interlock, and slope deposits widely cover fluvial sediments and the weathered or redeposited material. The main objective was to delimit the distribution of the fluvial deposits and thus to reconstruct the course of the pleistocene River Danube. The reconstruction of this layer configuration could only be done by comparison to adjacent seismic profiles. In total, eight overlapping profiles with geophone spacings of 3 and 5 m and numerous shotpoints provided

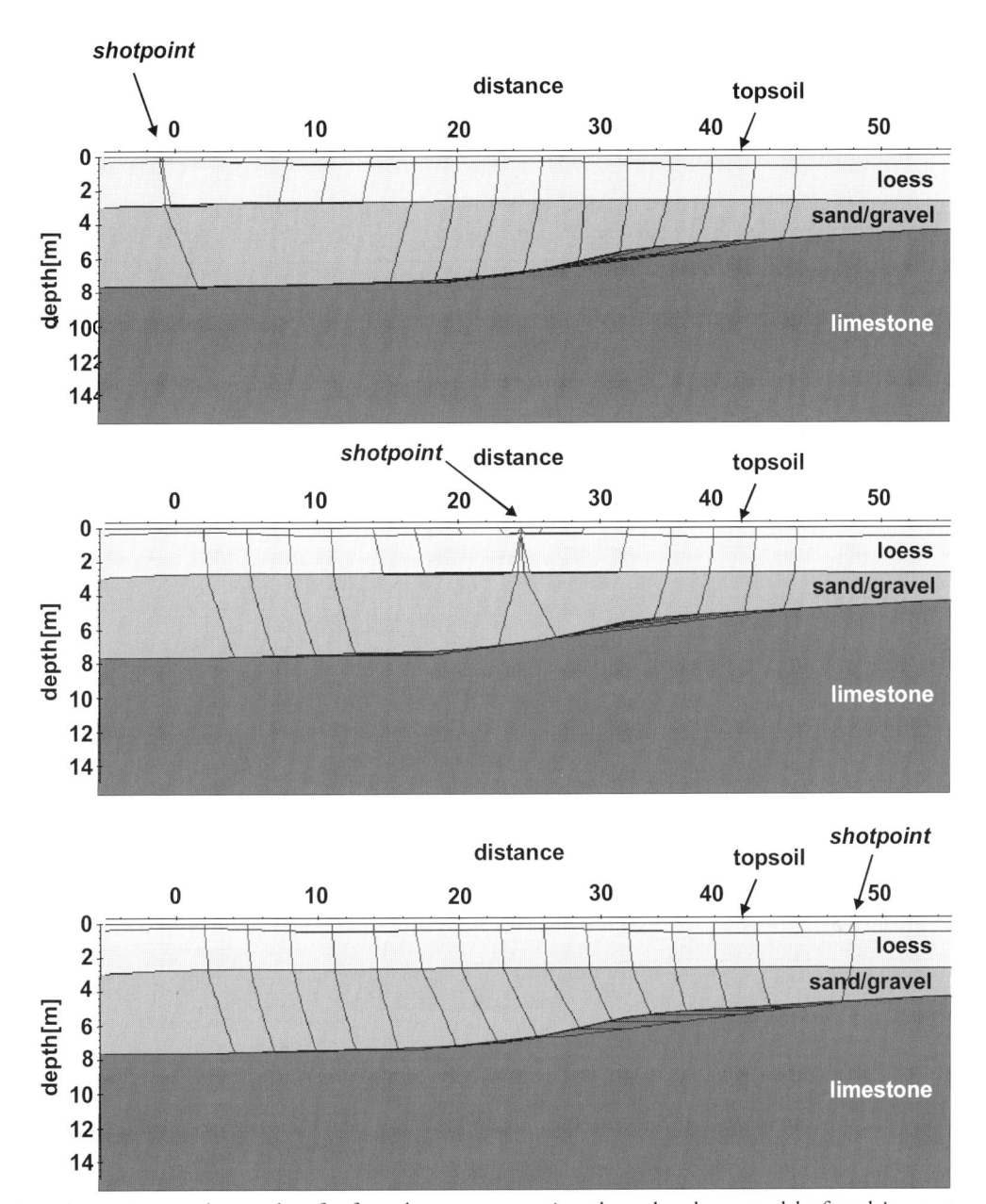

Fig. 4. a. Computed raypaths of refracted wave propagation through a layer model of a pleistocene terrace of the River 'Enz' near Bietigheim (Baden-Württemberg, Germany). Refracted waves appear at every interface so that traveltime information from all layers is available.
b. The same profile as in Figs. 4a and 4b but with a different shotpoint. Traveltime information from all layers is available, too.
c. The same profile as in Figs. 4a and 4b but with a different shotpoint. From the interface loess/sand/gravel no traveltime data is available. The refracted waves at the limestone interface outruns the waves refracted at the layer above. In this case, the sand/gravel appears as a hidden layer. Regarding all shotpoints (Figs. 4a–c), this layer is characterised as a 'semi-hidden layer' according to SJÖGREN (1984).

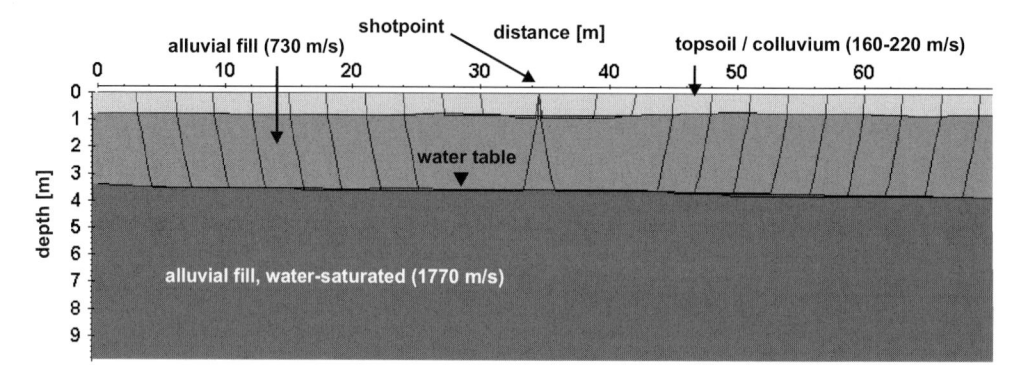

Fig. 5. Computed raypaths of refracted wave propagation through a layer model of alluvial fill in a valley of the 'Kraichgau' near Untergrombach (Baden-Württemberg, Germany). The loose, alluvial sediments beneath the groundwater table could not be further differentiated.

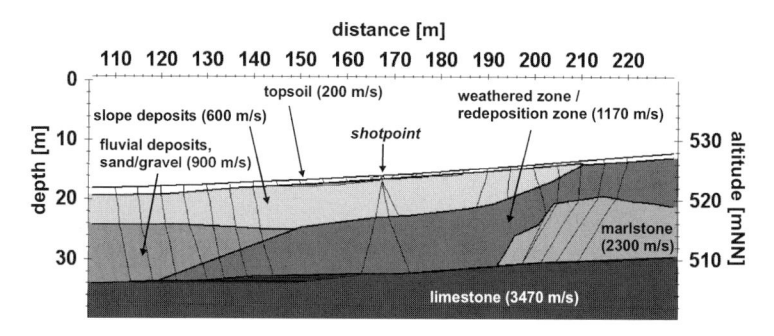

Fig. 6. Computed raypaths of refracted wave propagation through a layer model of a former valley of the River Danube which today appears as a dry valley. The location is situated near Ehingen/Donau (Baden-Württemberg, Germany). This part of a complete cross-section through the valley marks the transition from the valley bottom to the slope. The complex configuration of strata could only be resolved by comparison with the results of adjacent profiles.

the data modelling the subsurface layering. Accompanying drillings confirmed the results, mainly concerning the lateral distribution of strata. The lateral joint between slope deposits and the weathered or redeposited material – as shown in the example in Fig. 6 – could be reconstructed exactly. This example illustrates the strength of seismic refraction methods to resolve complex layer configurations. Drillings revealed data that are consistent with the results of the seismic measurements. The distribution of the fluvial deposits could be detected as well as the interlocking of slope deposits and a weathered zone/redeposition zone. Furthermore, changes in the underlying bedrock (limestone/marlstone) could be detected due to substantial changes in the seismic velocities.

The measurements used in the model displayed in Fig. 7 were taken on the opposite slope of the same valley as in Fig. 6. Here, the main objective was the detection of different types of

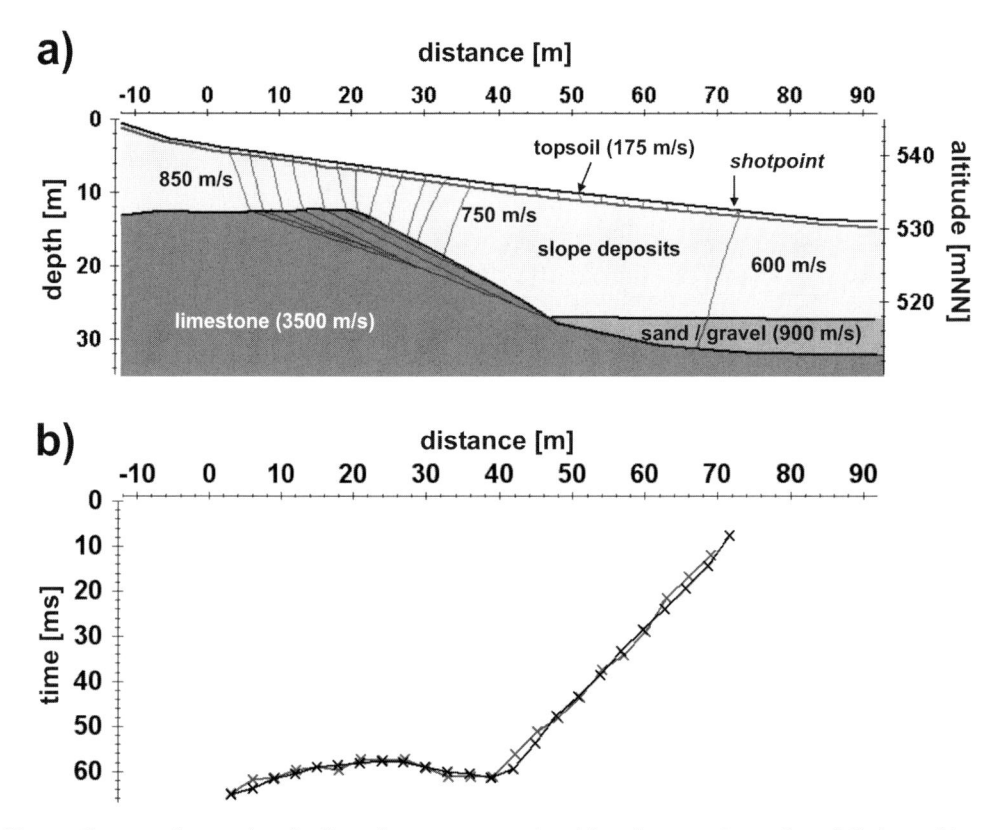

Fig. 7. Computed raypaths of refracted wave propagation (a) and comparison of modelled traveltimes with observed traveltimes (b). The measurements were taken in the same valley as shown in Fig. 6 but on the opposite slope. The step formed in the limestone basement is interpreted as a pleistocene undercut slope covered by slope deposits. The traveltime curve (b) sketches the topography of the subsurface limestone. The fluvial deposits (sand/gravel) could only be detected by comparing adjacent seismic profiles.

slope deposits and the reconstruction of the palaeosurface relating to the pleistocene River Danube. In this case, lateral changes in seismic velocities could be associated with different sediment types and the results were validated by drillings. The different seismic velocities of the slope deposits (600–850 m/s) are due to sediments with different grain size distribution and compaction. A higher proportion of clay and the absence of debris occurs in the upper part of the slope (850 m/s), whereas more debris and a higher silt content generated velocities of 600–750 m/s. The higher velocity is most likely due to older and therefore more weathered strata which had been conserved in this part of the slope. Fig. 8 shows a diagram of the travel time curves for selected shots that indicate the lateral changes in layering along the seismic profile.

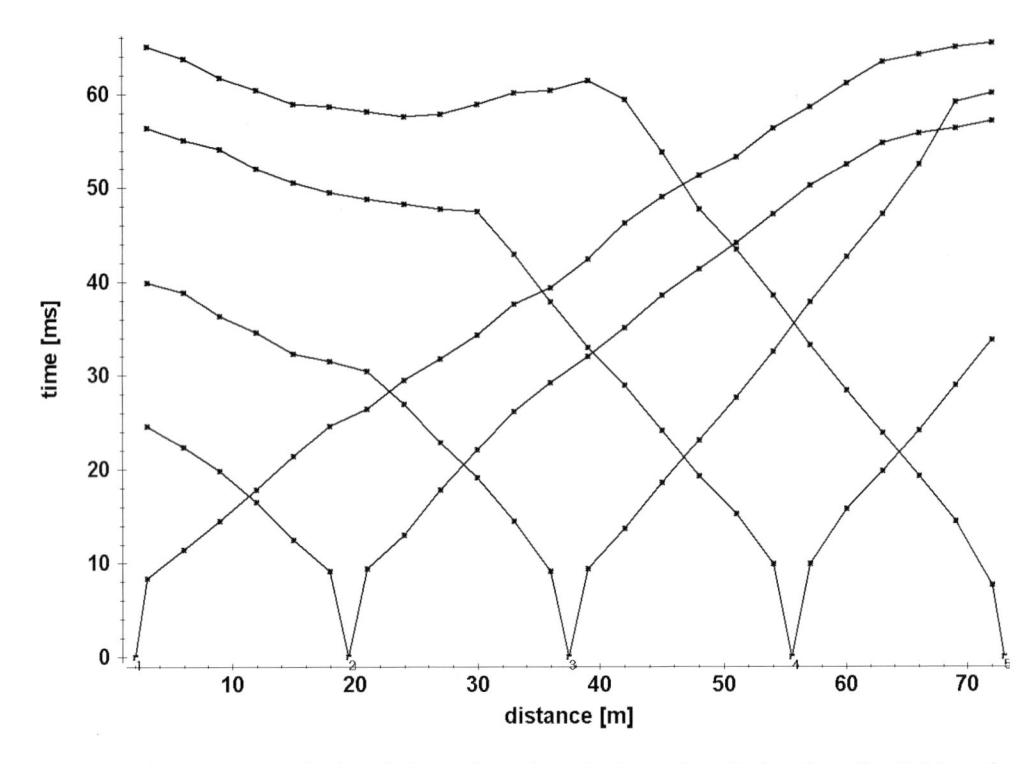

Fig. 8. Travel time curves of selected shots along the seismic section displayed in Fig. 7. Note the different travel time sections of forward and reverse shot that indicate the lateral changes in the layering displayed in Fig. 7.

2.2.2 Case study 'Maustobel': Investigations of a landslide area

The cross-section through a landslide area at the cuesta scarp of the 'Schwäbische Alb' near Eckwälden (Baden-Württemberg, Germany) in Fig. 9 is characterised by large differences in seismic velocities between the slide masses/slope deposits (250–600 m/s) and the underlying clay/claystone (2170 m/s). RAME GOWDA et al. (1998) report similar velocity differences between slide masses and basement in a case study located in the Himalayan foothills. Landslide areas are therefore suitable objects for investigation with seismic refraction methods. The lateral changes in the slide masses are probably caused by different levels of consolidation. The lowest velocity was measured at the steepest part of the slope. A further vertical differentiation of the slide masses/slope deposits was not possible due to the insufficient thickness of this layer. Even though results from drillings provided evidence for differences in grain size distribution or different proportions of debris of the slope deposits. The measurements were made using a geophone spacing of 2 m and several offset and inline shots. Fig. 10 shows the traveltime curves of selected shots along the profile. The outstanding variations between 30 and 40 m are due to different depths of the underlying bedrock, that occurs relatively near the surface in this part.

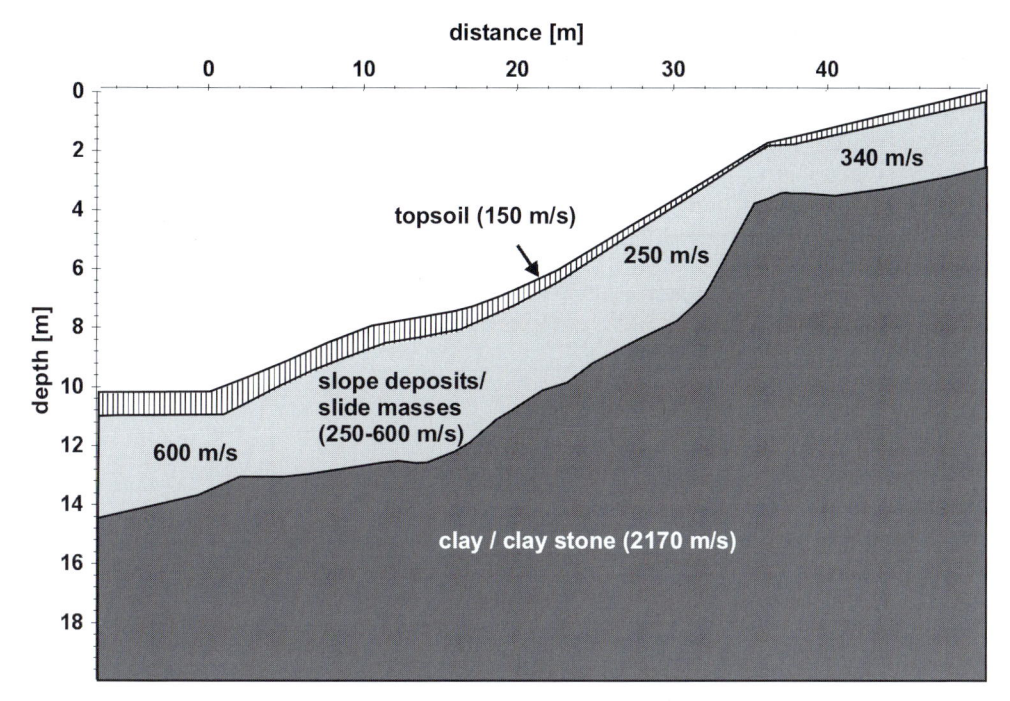

Fig. 9. Cross-section through a landslide area at the cuesta scarp of the 'Schwäbische Alb' (Baden-Württemberg, Germany) near Eckwälden. It was possible to differentiate the slope deposits/slide masses from clay/claystone because of the large difference in the seismic velocities. The results could be confirmed by drillings.

3 Differentiation of loose sediments from bedrock

The differentiation of loose sediments from bedrock is not problematic in most cases because of large differences in seismic velocities. Even if the determination of correct depths is difficult (problems concerning 'hidden layers', see above), lateral changes in layering or the topography of the bedrock can be reconstructed with seismic 2D-interpretation tools. Thus the results from seismic refraction measurements provide useful information about landscape evolution, e.g. the reconstruction of palaeosurfaces that are currently covered with sediment. In Fig. 7 the traveltime curve sketches the topography of the limestone which is assumed to be a former undercut slope of the pleistocene River Danube.

For a better illustration of the layering a 3D-display is shown in Fig. 11. This 3D model was generated by interpolating the results of three parallel seismic profiles measured on the pleistocene terrace introduced in Figs. 4a–c. The model shows lateral changes and the topography of the bedrock as well as the vertical differentiation of the loose sediments covering the underlying limestone.

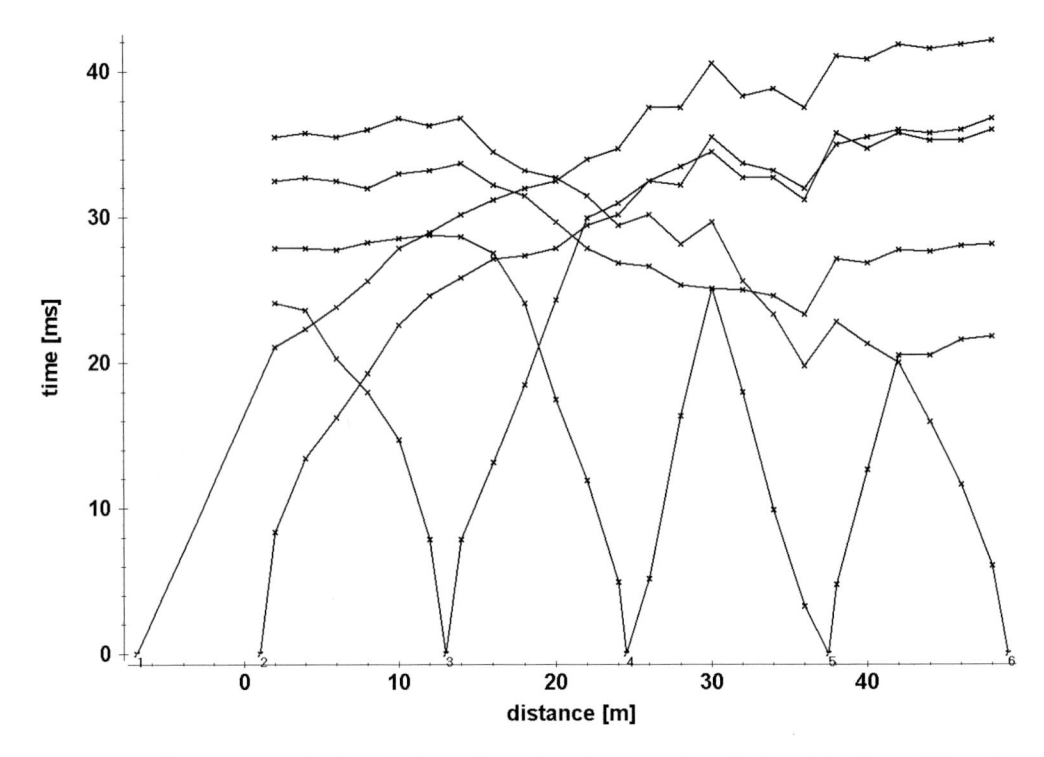

Fig. 10. Traveltime curves of selected shots along the seismic section displayed in Fig. 9. Note the different slopes of the travel time curves between 10 and 20 ms that indicate lateral changes of the slope desposits. The travel time variations between 30 and 40 m are due to different depths of the underlying bedrock (Fig. 9).

4 Conclusions

The vertical differentiation of strata with seismic refraction methods is sometimes affected by 'hidden layers' or by insufficient differences in the seismic velocities of different materials. The fluvial sediments of the case study 'Bietigheim' were only partly detectable with seismic refraction methods whereas the data of the case study 'Untergrombach' revealed the occurrence of a shallow ground water table that affected the interpretation of the layers beneath. Nevertheless, the 2D-methods that include raytracing techniques have the great advantage that lateral changes in layering can be detected with a high degree of accuracy. The results of the case study 'Kirchener Tal' showed clear velocitiy contrasts of different types of loose sediments. Thus it was possible to detect the sediment-covered pleistocene palaeosurface which is important for the reconstruction of the landscape evolution. Irregular refractors in solid bedrock can be mapped as well as lateral changes in loose rock. The delimitation of slide masses from bedrock (case study 'Maustobel') is important concerning geomorphological surveys in landslide areas, e.g. to estimate the amount of potentially displaceable material for risk management. Well directed drillings or additional geophysical surveys improve the results in any case and can provide increased certainty when modelling the shallow subsurface with seismic refraction methods.

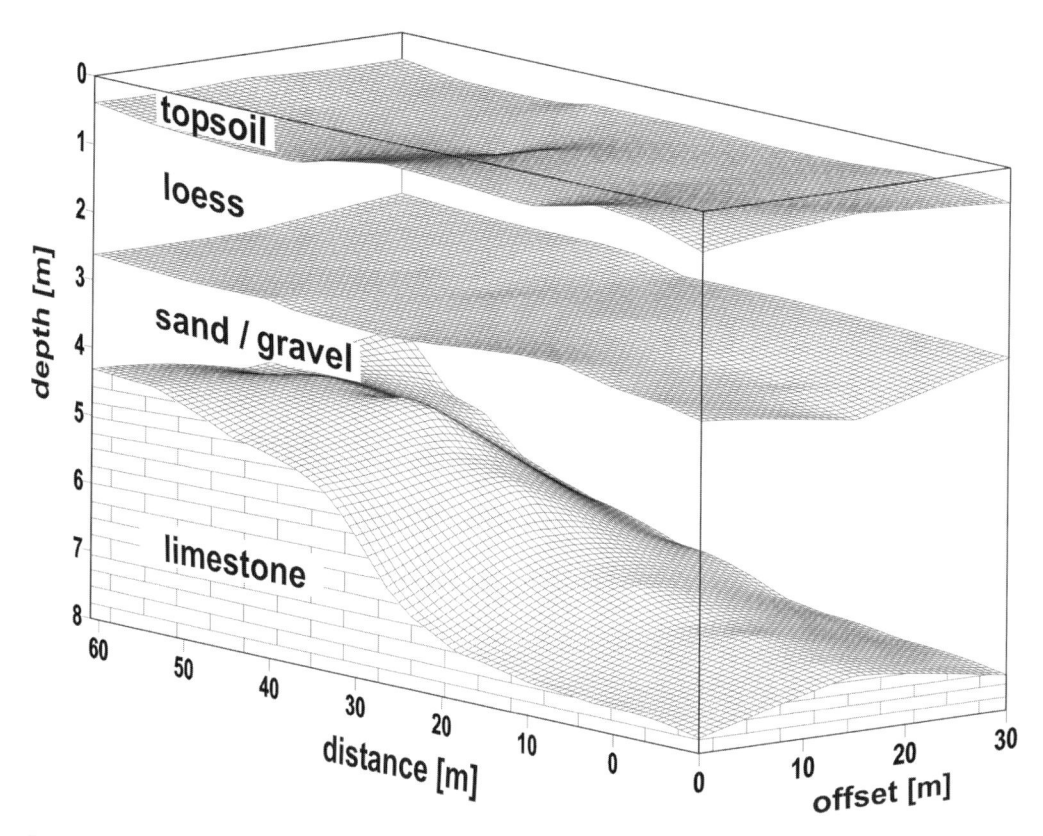

Fig. 11. 3D display of strata based on three parallel seismic profiles at a pleistocene terrace of the river 'Enz' near Bietigheim (Baden-Württemberg, Germany, same location as in Figs. 4a–c).

References

Burger, H. R.(1992): Exploration geophysics of the shallow subsurface. – 489 pp., Englewood Ciffs, New Jersey.

Eberle, J., Hecht, S. & P. Wittmann (2001): Neue Erkenntnisse zur Landschaftsgenese des Kirchener Tales bei Ehingen a. d. Donau mit Hilfe sedimentologischer und refraktionsseismischer Untersuchungsmethoden. – Jber. Mitt. Oberrhein. Geol. Ver., N.F., **83**: 339–355.

Fertig, J. (1997): Seismik – Prinzip der Methode. – In: Knödel, K., Krummel, H. & Lange, G. (Hrsg.): Handbuch zur Erkundung des Untergrundes von Deponien und Altlasten. Bd. 3: Geophysik. – S. 405–445, Berlin, Heidelberg.

Hecht, S. (2000): Fallbeispiele zur Anwendung refraktionsseismischer Methoden bei der Erkundung des oberflächennahen Untergrundes. – Z. Geomorph. N.F., Suppl.-Bd. **123**: 111–123.

Hecht, S. (2001): Anwendung refraktionsseismischer Methoden zur Erkundung des oberflächennahen Untergrundes. – Stuttgarter Geogr. Stud. **131**, 165 S.

Kirsch, R. & Rabbel, W. (1997): Seismische Verfahren in der Umweltgeophysik. – In: Beblo, M. (Hrsg.): Umweltgeophysik. – S. 243–311, Berlin.

Krammer, K. (1997): Untersuchung von Altlasten in ehemaligen Kiesgruben mit Hilfe der Refraktionsseismik. – In: Beblo, M. (Hrsg.): Umweltgeophysik. – S. 313–321, Berlin.

LANKSTON, R. W. (1989): The seismic refraction method: A viable tool for mapping shallow targets into the 1990 s. – Geophysics **54** (12): 1535–1542.

LANKSTON, R. W. (1990): High-Resolution Refraction Seismic Data Acquisition and Inter-pretation. – In: WARD, S. H. (Ed.): Geotechnical and environmental geophysics Vol. 1: Review and tutorial. – pp. 45–73, Society of Exploration Geophysicists, Tulsa.

MAVKO, G., MUKERJI, T. & DVORKIN, J. (1998): The rock physics handbook. Tools for seismic analysis in porous media. – 329 pp., Cambridge. New York.

PALMER, D. (1981): An introduction to the generalized reciprocal method of seismic refraction interpretation. – Geophysics **46** (11): 1508–1518.

RAME GOWDA, B. M., GHOSH, N., WADHWA, R. S., AKUT, P. V. & VAIDYA, S. D. (1998): Seismic refraction and electrical resistivity methods in landslide investigations in the Himalayan Foothills. – Environm. Engin. Geosci. **IV** (1): 130–135.

REYNOLDS, J. M. (1997): An introduction to applied and environmental geophysics. – 796 pp., Wiley, Chichester.

SANDMEIER, K.-J. (1997): Refraktionsseismik: Standard-Inversionsverfahren. – In: KNÖDEL, K., KRUMMEL, H. & LANGE, G. (Hrsg.): Handbuch zur Erkundung des Untergrundes von Deponien und Altlasten. Bd. 3: Geophysik. – S. 533–545, Berlin. Heidelberg.

SANDMEIER, K.-J. (2000): REFLEX – Version 2. Program for processing and interpretation of reflection and transmission data. – 235 pp., Karlsruhe.

SANDMEIER, K.-J. & LIEBHARDT, G. (1997): Refraktionsseismik: Iterative Interpretationsmethoden. – In: KNÖDEL, K., KRUMMEL, H. & LANGE. G. (Hrsg.): Handbuch zur Erkundung des Untergrundes von Deponien und Altlasten. Bd. 3: Geophysik. – S. 546–552, Berlin, Heidelberg.

SCHANZ, U. (1996): Geophysikalische Untersuchungen im Nahbereich eines Karstsystems (Westliche Schwäbische Alb). – Tübinger Geowiss. Arb. **C 29**, 114 S.

SHARMA, P. V. (1997): Environmental and engineering geophysics. – 475 pp., Cambridge.

SJÖGREN, B. (2000): A brief study of applications of the generalized reciprocal method and of some limitations of the method. – Geophys. Prospect. **48**: 815–834.

SJÖGREN, B. (1984): Shallow refraction seismics. – 268 pp., London, New York.

WHITELEY, R. J. (1992): Comment on 'The resolution of narrow low-velocity zones with the generalized reciprocal method'. – Geophys. Prospect. **40**: 925–931.

Address of the author: Dr. Stefan Hecht, Geographisches Institut der Universität Heidelberg, Im Neuenheimer Feld 348, D-69120 Heidelberg, Germany. E-mail: stefan.hecht@urz.uni-heidelberg.de

Use of Ground Penetrating Radar (GPR) soundings for investigating internal structures in rock glaciers. Examples from Prins Karls Forland, Svalbard

Ivar Berthling, Oslo, Bernd Etzelmüller, Blindern, Morgan Wåle, Vettre and Johan Ludvig Sollid, Blindern, Norway

with 5 figures and 4 tables

Summary. We present Ground Penetrating Radar (GPR) profiles from four rock glaciers on Prins Karls Forland, Svalbard. Using these profiles as examples, we discuss the ability of the GPR technique for revealing information about internal structures in rock glaciers. Our GPR profiles were collected with a RAMAC GPR system, equipped with 50 MHz antennas. The maximum penetration depth obtained was about 30 m. In most cases, the reflectors visible on the profiles correspond to reflections from material boundaries or thin layers within the rock glaciers. We also show examples of how surface reflections are displayed on the radar profiles. Reflectors could often be matched between longitudinal and transverse profiles at their intersections. The profiles reveal reflectors that have a common structural development on the longitudinal profiles while the transverse profiles have variably dipping or wavy reflectors with less comparable appearance from profile to profile. In the lower parts, some profiles show one or two zones of more evenly developed reflectors that may be traced along larger parts or the whole of the profile. On the two smaller investigated rock glaciers, these reflectors may correspond both to a shear zone and to the bedrock interface. On the two larger rock glaciers, the bedrock is mainly too deep to be detected and these reflectors may possibly be interpreted as a shear zone.

Zusammenfassung. In diesem Aufsatz präsentieren wir Bodenradarprofile (Ground Penetration Radar – GPR) von vier Blockgletschern, lokalisiert auf Prins Karls Forland, Svalbard. Mit Hilfe dieser Profile diskutieren wir die Möglichkeiten der Bodenradartechnik, Informationen über die interne Struktur von Blockgletschern zu erfassen. Unsere Bodenradarprofile wurden mit einem RAMAC GPR System, bestückt mit einer 50 MHz Antenne, aufgenommen. Die maximal gemessene Penetrationstiefe war ca. 30 m. In den meisten Fällen korrespondierten die Reflektoren, die sichtbar in den Profilen waren, mit Reflektoren von Materialgrenzen oder anderen dünnen Schichten in den Blockgletschern. Wir zeigen auch Beispiele, wie die Oberflächenreflektoren in den Profilen abgebildet werden. Die Reflektoren passten oft sehr gut in den Berührungspunkten zwischen Längs- und Querprofilen zusammen. Die Reflektoren in den Längsprofilen zeigten normalerweise eine gemeinsame strukturelle Entwicklung, während die Querprofile oft geneigte oder wellenartige Reflektoren aufzeigen, die weniger Gemeinsamkeiten zwischen den Querprofilen vermuten lassen. In den unteren Teilen der Blockgletscher zeigten einige der Profile Zonen mit mehr zusammenhängenden Reflektoren. Bei den kleineren der untersuchten Blockgletschern dürften diese Reflektoren mit einer Scherzone oder dem Felsuntergrund korrespondieren. Bei den zwei grössten Blockgletschern liegt der Übergang zum Fels zu tief, um mit unserer Ausrüstung identifiziert werden zu können. Dort sind die Reflektoren möglicherweise als Scherzone zu interpretieren.

Introduction

Ground Penetrating Radar (GPR) is a geophysical method for subsurface investigation that utilises electromagnetic signals transmitted into the ground as pulses from an antenna. A receiver antenna picks up energy that is partially reflected as the signal passes through a dielectric boundary in the ground. Utilisation of this technology started early in the 20th century (DANIELS et al. 1988), but commercial GPR systems have only been available since the mid 1970's. The first digitally controlled GPR system was introduced by Sensors and Software in the mid 1980's. Compared to other geophysical methods, GPR supplies data with very high vertical resolution, a potential high recording speed and real-time display of the acquired data. This enables mapping of structural features in the subsurface in 2D or 3D over relatively large areas.

Rock glaciers are creeping permafrost bodies. The material properties of rock glaciers are known fairly well from inspection of outcrops (e.g. BARSCH 1996), and during the last decades geophysical investigations and a few core drillings (e.g. HAEBERLI & VONDER MÜHLL 1996, VONDER MÜHLL & HOLUB 1992, ARENSON et al. 2002). The spatial distribution of structures and layers within a rock glacier is still, however, very little known. Several aspects of rock glacier research could benefit from improved knowledge in this field. This regards for instance the processes involved in material (debris and ice) accumulation on the rock glacier, the proposed continuity of a talus cone – rock glacier system (HAEBERLI et al. 1998), rock glacier 2D dynamics (detection of shear zone and patterns of layer displacement) and ultimately rock glacier evolution (including assessment of rock glacier age and time dependant or average rates of rock wall retreat). Possibly, such knowledge could also help differentiating relic rock glaciers from rock avalanche deposits. The GPR method offers the potential for retrieving the necessary information on structure distribution within the rock glaciers.

Although already employed by HAEBERLI et al. (1982), KING et al. (1987) and VONDER MÜHLL (1993) with some success, convincing results regarding the potential of the GPR technique on rock glaciers was first presented by VONDER MÜHLL & HUGGENBERGER (1997) and VONDER MÜHLL et al. (2000) from the Murtél-Corvatsch rock glacier. BERTHLING et al. (2000) and ISAKSEN et al. (2000a) used the technique on high arctic rock glaciers on Svalbard, showing that a partly layered structure could be traced along the longitudinal profile of these rock glaciers. GPR data from rock glaciers are currently also available from the USA (DEGENHARDT et al. 2002).

In the present study, we discuss GPR longitudinal and transverse profiles from four rock glaciers on Prins Karls Forland, Svalbard. All longitudinal profiles and one transverse profile have earlier been presented by BERTHLING et al. (2000), while additionally eight transverse profiles are presented here. Using these profiles as examples, we discuss the ability, limitations and potential pitfalls of the GPR technique with respect to revealing information about the internal composition of rock glaciers.

Setting

The study site is the north-western coastal area of Prins Karls Forland (78°50'N, 10°30'E), the westernmost island of the Svalbard archipelago (Fig. 1). Permafrost is continuous on Svalbard, except partly beneath glaciers, reaching depths of several hundred meters in the inland (LIESTØL 1976, ISAKSEN et al. 2000b). At the study site, rock glaciers form a several kilometer long continuous transition between the backing talus slopes and rock walls, and a strandflat area close to the present shoreline (Fig. 1). The up to 500 meter high backing mountains are dissected by large chutes, which probably contribute to most of the material supply to the rock glaciers. The setting provides fairly simple boundary conditions, with respect to topography, which ease the interpretation of the GPR profiles.

GPR system parameters and field procedures

Fieldwork for the study was carried out in August 1998. We used a RAMAC GPR system from Malå GeoScience equipped with 50 MHz antennas. Transmitter and receiver antennas were fixed at one meter distance, with antennas orientated transverse to the profile direction, and linked to the controlling unit and a computer by fibre-optical cables. Due to the coarse surface of the rock glaciers, the antennas were carried along the profiles about 0.5 m above the surface. Three persons were required to operate this set-up in the field. The electromagnetic pulses were triggered at a fixed distance of 0.5 m, controlled by a hip-chain. The sampling frequency should be high enough to avoid aliasing. We used 500 samples and a sampling frequency of 481.06 MHz, which gives and a time window of 1.04 μs. The signal to noise ratio was improved by stacking. In the stacking procedure, a number of measurements (in this case 16) are performed at each point, and the mean values are stored.

Two transverse profiles were collected on rock glaciers number 7, 12 and 15, and three on rock glacier number 9. The profiles were located close to the front and at the break of slope between the rock glacier and the backing talus slope. The upper transverse profile on rock glacier 9 was located on the talus cone. Further, one longitudinal profile was collected on each of these rock glaciers. The antennas in our study were oriented normal to the profile direction, and thus the electrical field was differently polarised relative to the layers on the longitudinal and transverse profiles. A Common Mid Point (CMP) measurement was done on rock glacier number 12. Positioning of the profiles were done through tying them to existing displacement measurement target points or large boulders that could be identified on air photographs.

All data processing was performed using the software GRADIX (©Interpex Ltd). Due to the rapid attenuation of the radar signals, it is necessary to apply a time dependent gain function. It was important to show stratigraphic horizon continuity, and we therefore applied an

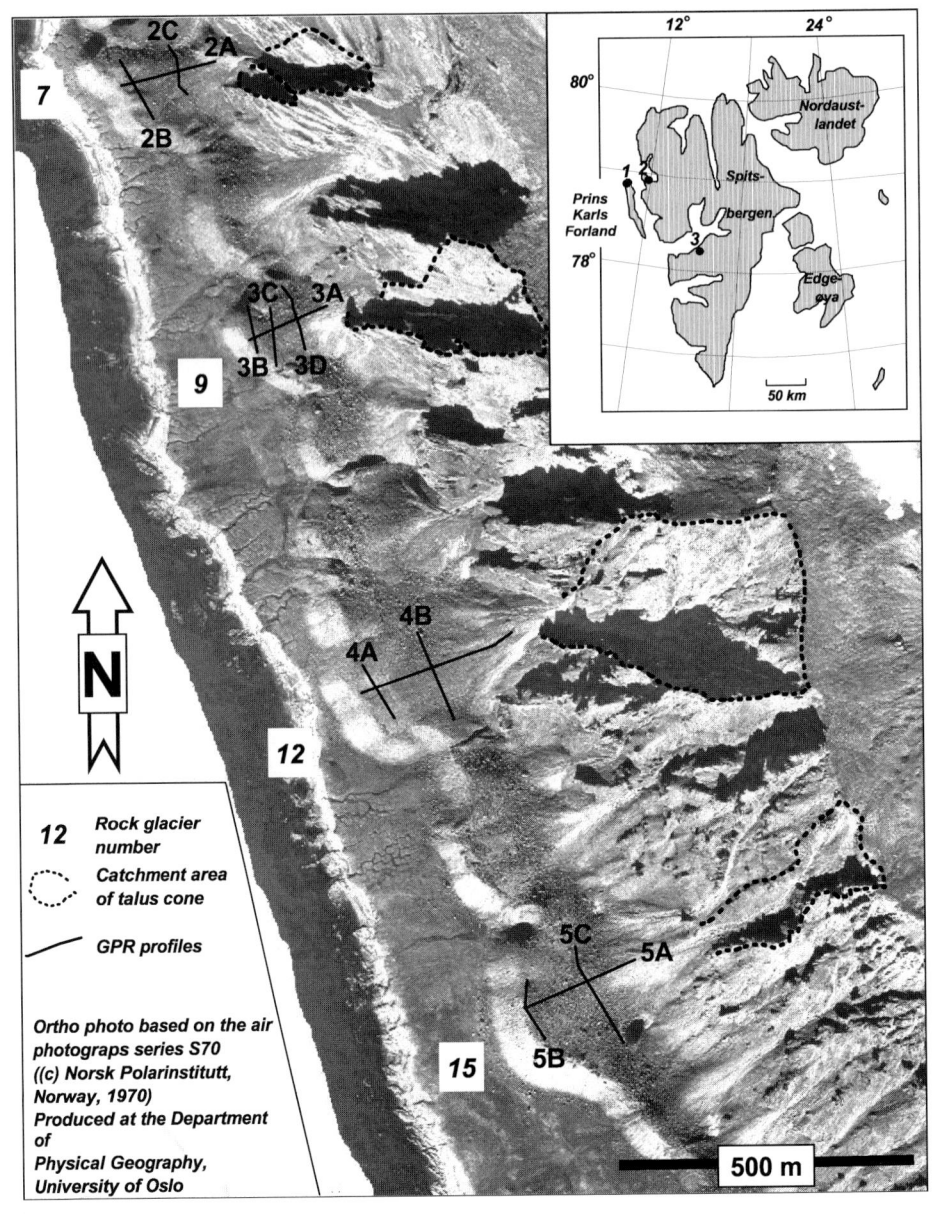

Fig. 1. Key map in the upper right shows the location of *1* Fuglehuken on the northernmost part of Prins Karls Forland, *2* Ny-Ålesund and *3* Longyearbyen. Orthophoto of the study site at Fuglehuken, Prins Karls Forland. Main figure is an ortophoto of the study site, showing the location of the investigated rock glaciers 7, 9, 12 and 15. The GPR profiles in solid lines are numbered according to the figure number where they are displayed. Dotted lines show approximate catchment area of talus cones from which the investigated rock glaciers develop. Dark shadings are due to shadows, except for a small pond on the south-eastern side of rock glacier 15 talus slope and the Greenland ocean on the extreme western part of the orthophoto. The coastline and the main parts of the rock glacier fronts are in light shadings.

Automatic Gain Control (AGC). This gain function amplifies (or attenuates) the data point at the centre of a time window by the ratio of the desired output value to the average signal amplitude within the time window. Amplitude fidelity is thus not maintained. The gain window applied to each profile (Figs. 2–5) varies slightly. The results from the CMP profile provided a basis for depth migration of the profiles, which then were corrected for topography (BERTHLING et al. 2000). Topographic data were obtained from a Digital Elevation Model (DEM) (BERTHLING et al. 1998).

Results

The Common Mid Point (CMP) measurement

For the main parts of the profile, the radiowave velocity of the medium was 0.14 m ns^{-1}, while the active layer had a velocity of 0.055 m ns^{-1}. This corresponds to a dielectric constant for the permafrost of 4.6. Although subsurface layers are not horizontal in this area, partly violating requirements of the CMP method, the obtained velocity is reasonable. Our obtained velocity was also used by ISAKSEN et al. (2000a) and is well in line with LEHMANN et al. (1998) who assume a value of 0.15 m ns^{-1} on the Murtèl-Corvatsch rock glacier.

The GPR transverse profiles

The position of the GPR profiles on Prins Karls Forland is shown in Fig. 1. The profiles are displayed in Figs. 2–5. As the radiowave velocity of the active layer is much lower than that of the permafrost (about 1/3), the active layer is displayed almost three times thicker than "real" on the radar profiles. Depth to any reflector in the presented GPR profiles therefore refers to the scale displayed in Figs. 2–5 ("apparent depth" and not to "true" depth). The accuracy of the DEM used for terrain correction and the matching of the profiles to the DEM also introduce errors. Absolute heights of reflectors are therefore referred to as apparent height above sea level (AHASL) and contain both types of errors. We focus on the internal structures in the rock glacier, and the active layer is thus not considered any further. Tables 1–4 summarise reflector development on the transverse profiles, and depths to reflectors visible on transverse and longitudinal profile intersections. As no markers were applied to the GPR profiles in the field at the intersections of the longitudinal and transverse profiles, the intersections were found from the outline of the profiles drawn on air photographs. The apparent depths to reflectors at the intersections shown in Tables 1–4 can therefore not be expected to match exactly. In retrospect, simultaneous profiling with a differential GPS receiver would have been preferable.

A common feature of most transverse profiles is that the rock glacier can be divided into four zones with respect to the type of the reflector patterns. The active layer represents the upper zone, where sharp and continuous reflectors are found. Below the active layer, there is a zone where no or only few/vague reflectors are found. Typical examples are found on rock glacier 7 (*1* in Figs. 2 B,C). This zone is also found on the longitudinal profiles (BERTHLING et al. 2000). Beneath, one enters a zone of scattered reflectors. These may be well devel-

Table 1. Description of the transverse profiles of rock glacier 7. Numbers in italics refer to pointers in Fig. 2. AHASL = apparent height above sea level, app. depth = apparent depth. In the columns L. P. (longitudinal profile) and T. P. (transverse profiles), we compare thickness of layers and depth to reflectors measured on the radar profiles at the profile intersections. Numbers in parenthesis: vague reflector. All values given to nearest half metre.

Fig. 2B – Profile T 7–1	L. P.	T. P.
Zone of no reflectors	Thickness (m)	
1: Below active layer to max 10 m apparent depth. Thickness 5 to 7.5 m	6.5	6.5
Reflectors/zones of reflectors	App. depth (m)	
2: Scattered reflection horizons, sub-parallel to surface.	11.0	11.5
	12.5	13.5
3: Pronounced reflector, dipping into rock glacier from profile end, AHASL 17.5–22 m		22–27

Fig. 2C – Profile T 7–2	L. P.	T. P.
Zone of no reflectors	Thickness (m)	
1: Below active layer, down to max. 11 m app. depth. 6 to 9 m thickness.	9.0	9.0
Reflectors/zones of reflectors	App. depth (m)	
	12.0	
2: Scattered, wavy reflections	13.0	13.0
		14.0
3: System of reflectors, continuous along parts of the profile, AHASL 20–35 m	16.0	18.0
4: Vague reflectors along parts of the profile, AHASL 24 m		

oped, but seldom along great distance. Typical length of these reflectors are 10 m. These reflectors are mainly dipping or wavy. A general impression is that they are sub-parallel to surface slope, especially on the profiles on or near the talus cones (e.g. *2* in Fig. 3D and *2* in Fig. 4B). However, there are also examples of the opposite, as on rock glacier 9 and 12 (*2* in Fig. 3C; *2* in Fig. 4A). The apparent depths of these reflectors often match well, mainly within a metre, with depths to reflectors on the longitudinal profile, at the respective intersections (Tables 1–4). The lower zone consists either of a set of reflectors or a sharp reflector, developed more or less continuously across the profile. The best examples are from rock glacier 7 and 9 (*1* in Fig. 2A; *3* in Fig. 3D). In addition, some deeper reflectors are found on some of the profiles (e.g. *1* in Fig. 3A).

Discussion

Depth penetration of the GPR system

The rock glaciers on Prins Karls Forland are up to about 50 m thick in their frontal areas. The depth penetration of the GPR system should therefore ideally be in that range in order to detect both bottom topography and possible reflectors within the rock glacier. The depth

Table 2. Description of the transverse profiles of rock glacier 9. Numbers in italics refer to pointers in Fig. 3. AHASL = apparent height above sea level, app. depth = apparent depth. In the columns L. P. (longitudinal profile) and T. P. (transverse profiles), we compare thickness of layers and depth to reflectors measured on the radar profiles at the profile intersections. Numbers in parenthesis: vague reflector. All values given to nearest half metre.

Fig. 3B – Profile T 9–1	L. P.	T.P
Zone of no reflectors	Thickness (m)	
1: Below active layer, down to max 12.5 m apparent depth. 6 to 8 m thickness.	5.5	7.0
Reflectors/zones of reflectors	App. depth (m)	
2: Mainly well developed reflectors, continuous along parts of the profile. AHASL 23–35 m.	(10.0)	(10.0)
	(13.0)	12.5
	15.0	14.0
	17.0	16.5
3: Prominent reflector with an AHASl of 27 m and apparent depth of 16.7 m.	18.5	18.5
	20.0	
	21.5	20.5
	25.0	(24.0)
4: More or less continuous reflector. AHASL 11–19 m	(30.0)	31.0

Fig. 3C – Profile T 9–2	L. P.	T. P.
Zone of no reflectors	Thickness (m)	
1: Not continuously developed. Max apparent depth 12 m. Thickness 0–9 m.	6.0	4.5
Reflectors/zones of reflectors	App. depth (m)	
		8.5
	10.5	10.5
2: Well developed to vague, dipping reflectors, AHASL 30–39 m	14.0	12.5
	16.0	16.0
	17.5	18.0
3: Well developed reflector(s), but not continuous, AHASL 23–30 m	20.0	19.5
4: Evenly dipping reflector, AHASL 15–20 m		

Fig. 3D – Profile T 9–3	L. P.	T. P.
Zone of no reflectors		
1: Not evenly developed. Max apparent depth 13 m. Thickness 0–10 m.		
Reflectors/zones of reflectors	App. depth (m)	
2: Reflectors mainly dip parallel with talus cone, AHASL 30–50 m.	4.5	4.0
	6.0	
	11.0	11.0
	16.0	15.5
	27.0	28.0
3: Very well developed continuous reflector, AHASL 25–30 m.		
4: Umbrella structures from surface reflections, corresponding to linear slanting reflections (1) in Fig. 4A.		

Table 3. Description of the transverse reflectors within rock glacier 12. Numbers in italics refer to pointers in Fig. 4. AHASL = apparent height above sea level. The length profile was could not be migrated, and comparisons of reflectors at the profile intersections was therefore not possible.

Fig. 4A – Profile T 12–1	L. P.	T. P.
Zone of no reflectors *1*: Not continuously or well developed. Max thickness 5 m. *Reflectors/zones of reflectors* *2*: Well developed reflectors, with somewhat wavy appearance. *3*: Somewhat vague reflectors with a more even development along the profile. AHASL = 33–45 m		Unmigrated longitudinal profile; no comparisons done.

Fig. 4 B – Profile T 12–2	L. P.	T. P.
Zone of no reflectors *1*: Vague reflectors are found, but these seem to be multiples from the active layer. Max thickness of zone is 6 m. *Reflectors/zones of reflectors* *2*: Well developed reflectors, with somewhat wavy appearance. A general tendency for layers to dip along surface slope to each side of the talus cone centreline. AHASL 25–31 m		Unmigrated longitudinal profile; no comparisons done.

penetration of a radar system is, however, not straight forward to predict. A basic requirement is that the received reflected power from an object must be strong enough for the system to detect it. This can be evaluated by the radar range equation, which relates the received power from a scattering object to the transmitting power, antenna gain and the distance to the object, and by the signal to noise ratio of the receiver. However, many of the parameters in the radar equation are generally not known. GPR system characteristics such as performance factor and antenna pattern provide basic constraints, while the electrical properties of the ground, character and size of the reflector in question are site dependent constraints. Therefore, the basic decision to be made is that of antenna centre frequency. Lower frequencies generally give better depth penetration, but there is a trade off on resolution and portability. Forward modelling, offered by some commercial GPR interpretation software packages, is a useful tool to aid in this choice.

The maximum apparent penetration depth encountered in our investigations was nearly 30 m. The prominent reflection at about 27 m depth, *3* in Fig. 3D is the best example. The deeper reflectors in this (*4* in Fig. 3D) and other profiles are probably surface reflections of waves propagating in the air (discussed below).

The centre frequency should also be chosen so as to reduce clutter, the signal returns from material heterogeneity in soils and rock (ANNAN 1998). In a rock glacier, the materials encountered may range from large boulders and air voids, through fine debris saturated or supersaturated with ice, to more or less clean ice layers. The typical clutter dimension should by much shorter that the signal wavelength, that is much shorter than V/f, where V is the velocity of the medium and f is centre frequency. ANNAN (1998) states that a factor of 10 between signal

Table 4. Description of the transverse profiles of rock glacier 15. Numbers in italics refer to pointers in Fig. 5. AHASL = apparent height above sea level, app. depth = apparent depth. In the columns L. P. (longitudinal profile) and T.P. (transverse profiles), we compare thickness of layers and depth to reflectors measured on the radar profiles at the profile intersections. Numbers in parenthesis: vague reflector. All values given to nearest half metre.

Fig. 5 B – Profile T 15–1	L. P.	T. P.
Zone of no reflectors	Thickness (m)	
1: Below active layer, max apparent depth 10 m. Thickness 5–10 m.	7.5	8.0
Reflectors/zone of reflectors	App. depth (m)	
2: Well developed reflectors, with somewhat wavy appearance. A general tendency for layers to dip along surface slope to each side of the talus cone centreline.	11.0	11.0
	13.5	13.0
	16.5	16.0
	17.5	18.5
	20.5	20.5
	22.5	22.5
3: Zone of relatively horizontal reflectors, developed across entire profile. AHASL 40–46 m.	23.0	23.0
	24.5	25.0
	26.0	25.5
4: Single reflector, somewhat separated from zone *3*.		27.5

Fig. 5C – Profile T 15–2	L. P.	T. P.
Zone of no reflectors		
1: Deeper than on other profiles, probably due to a 100–150 ns shorter AGC-window on this profile.	No matches, most likely due to differences in the AGC windows applied	
Reflectors/zones of reflectors		
2: Well developed reflectors on the southern part, vaguely continuing into central part of profile.		

wavelength and clutter dimension is appropriate. On Prins Karls Forland, the signal wavelength was about 1.1 m in the active layer for a 50 MHz antenna, and large clutter effects can thus be anticipated from the large stones and boulders found here. A 25 MHz antenna would have improved, but not eliminated, this effect. In the permafrost, the signal wavelength was about 3 m for the 50 MHz antenna and clutter effects are thus less probable.

Resolution

The resolution in GPR soundings is another fundamental issue. Both the horizontal and the vertical resolution must be considered. The horizontal resolution is determined by wavelength and depth, so that lower frequencies and larger depths decrease the resolution. Reflectors must have a radius larger than the first Fresnel zone in order to be resolved. With a 50 MHz antenna and a medium velocity of 0.15 m ns^{-1}, the wavelength is about 3 m. Minimum reflector radius f can be estimated from $f = \sqrt{\lambda h/2 + \lambda^2/16}$, where λ is wavelength and h

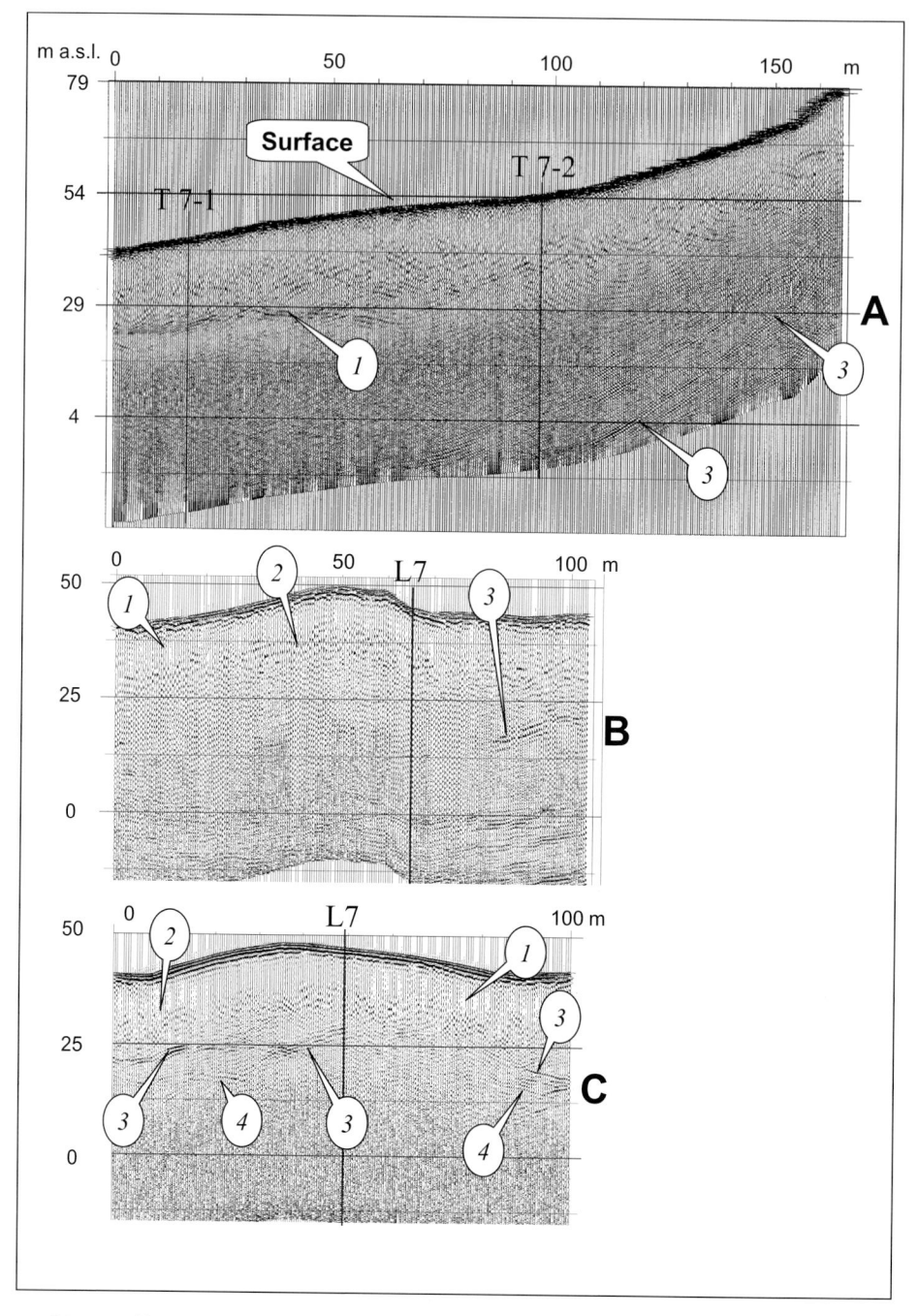

Fig. 2. GPR profiles on rock glacier 7. Horizontal and vertical scales are approximately equal on all profiles. Numbered pointers are referred to in Table 1 and in the text. **A**: longitudinal profile, **B**: frontal transverse profile, **C** transverse profile at rock glacier back.

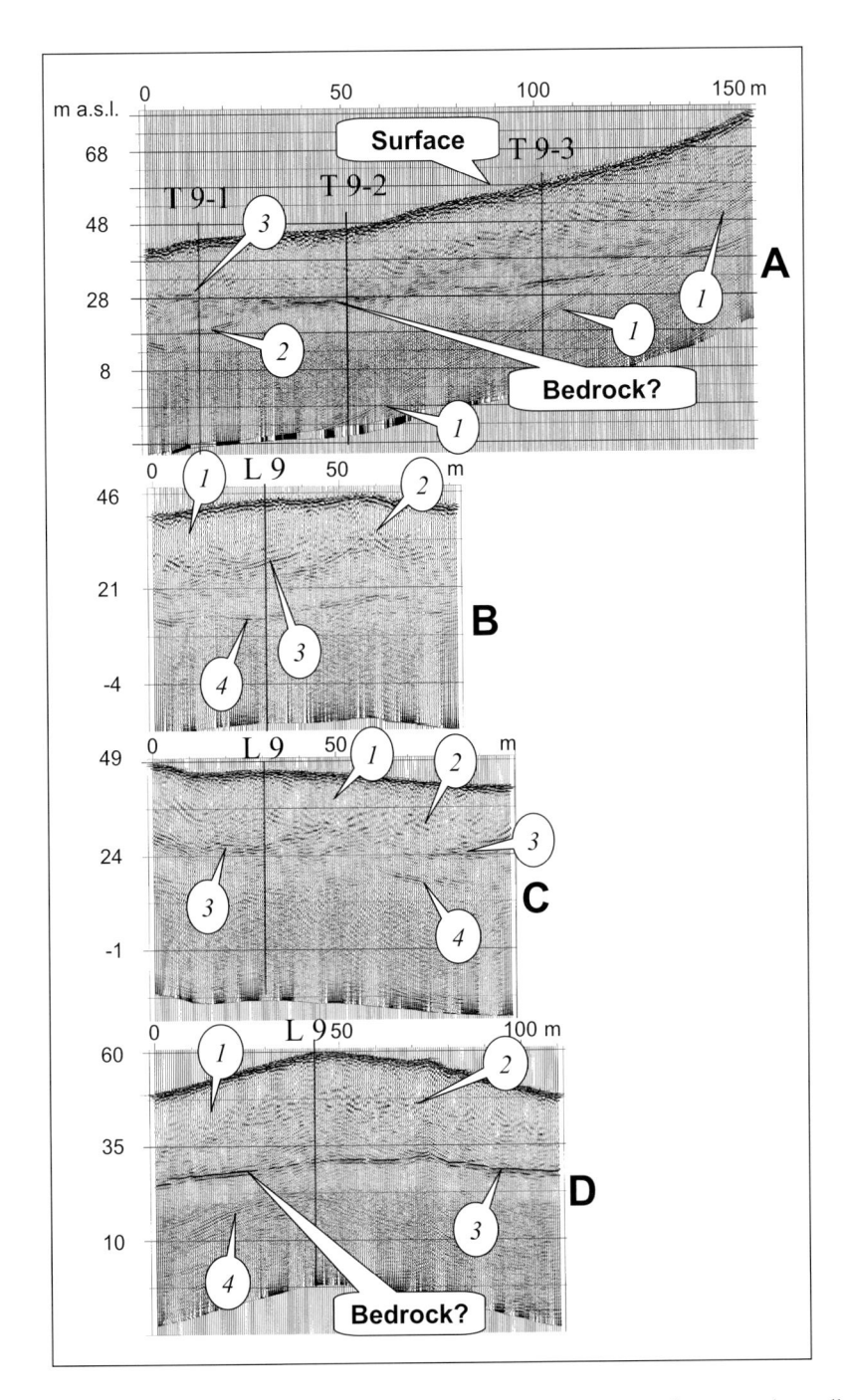

Fig. 3. GPR profiles on rock glacier 9. Horizontal and vertical scales are about equal on all profiles. Numbered pointers are referred to in Table 2 and in the text. **A**: longitudinal profile, **B**: frontal transverse profile, **C** transverse profile at rock glacier back, **D**: talus slope transverse profile.

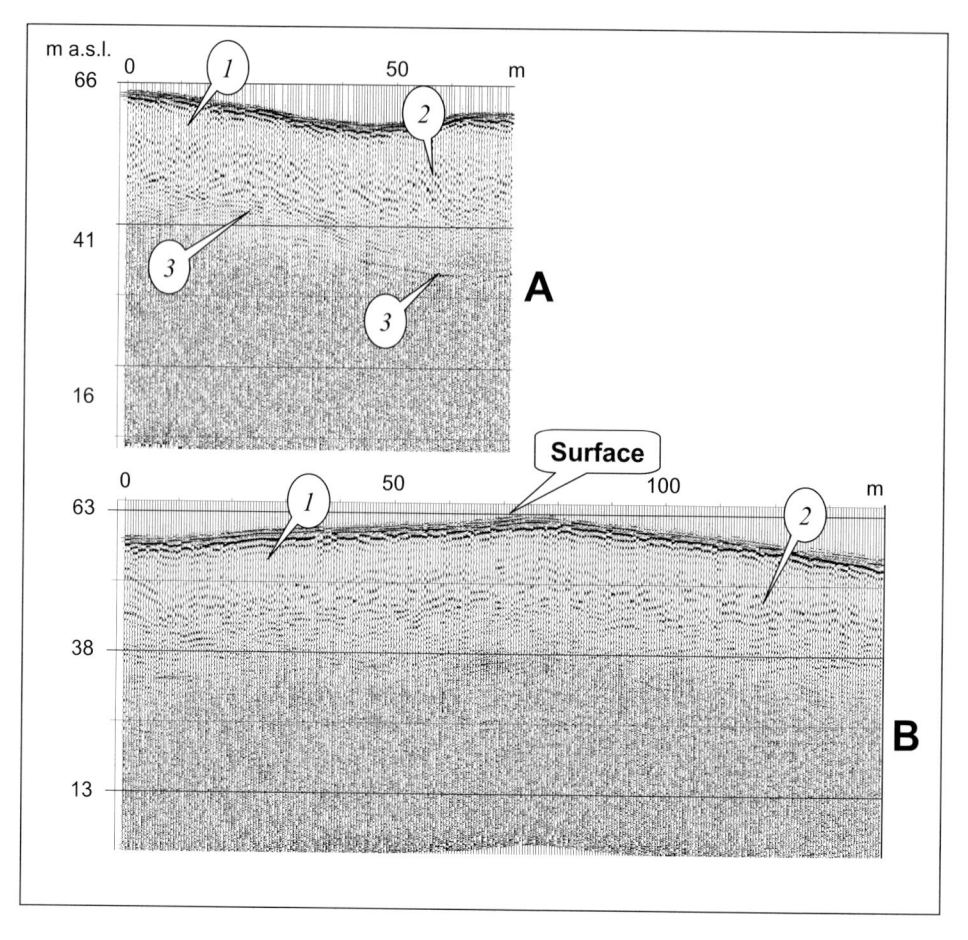

Fig. 4. GPR transverse profiles on rock glacier 12. Horizontal and vertical scales are about equal on both profiles. Numbered pointers are referred to in Table 3 and in the text. Longitudinal profile could not be migrated, and is not displayed. **A**: frontal transverse profile, **B** transverse profile at rock glacier back.

is depth to reflector (Mc Quillin et al. 1984). This yields a critical radius of 5.5 m at 20 m depth. Vertical resolution is the minimum distance between two reflectors so that these reflectors can be distinguished, and is determined by the wavelength and width of the reflected pulse. For a short pulse, true resolution is in the order of $\lambda/3-\lambda/2$ (Trabant 1984), and lower frequencies thus give lower resolution. On the rock glaciers, with a dielectric constant of 4.6 and an antenna centre frequency of 50 MHz, the resolution is not better than about one meter. In the active layer, we calculated a dielectric constant of 30 which gives a much better resolution (0.4–0.5 m). A high loss from scattering in the active layer cause more attenuation of the higher frequencies of the bandwidth. In the present case, any layer should be at least one meter thick if both its top and bottom is to be distinguished. However, even though an object is thinner than this, it can still give strong reflections.

Sources of reflections

As the electromagnetic waves propagate into the ground, reflections are generated from the boundaries of materials of different electrical properties. For wave propagation normal to a boundary between sufficiently thick layers with minor contrasts in conductivity and magnetic permeability, the reflection coefficient is $R \approx ((K_1)^{1/2} - (K_2)^{1/2})/((K_1)^{1/2} + (K_2)^{1/2})$, where K_1 is the dielectric constant of the host media and K_2 the dielectric constant of the target (layer) media. For reflections to be identified, a rule of thumb states that R^2 should be larger than 0.01 (ANNAN & COSWAY 1992).

Beneath the active layer on rock glaciers, the interfaces encountered are between ice saturated debris layers, ice layers, possibly porous debris layers and bedrock. According to tabulated values in textbooks and manuals, the dielectric contrasts between such materials are not very large. They are nevertheless often large enough to cause reflections, given the rule of thumb stated above, although this is difficult to asses in the case of the rock glacier/bedrock interface. In general, the lack of large dielectric contrasts within the rock glaciers, once the active layer is penetrated, is probably decisive for the great penetration depth achieved.

If the wavelength of the electromagnetic waves is large relative to the thickness of a layer, the situation is more complicated. Generally, the amplitude of the reflection will depend both on the thickness of the layer and the Fresnel reflection coefficient of the media, and the high frequency energy will be reflected while the lower frequencies are transmitted (ANNAN 1998). Another potential problem is that although one most often treats reflections as caused by wave propagation through layers of different materials, they could originate from anisotropy due to fine layering of the material. For instance, BRISTOW et al. (2000) use GPR to map sedimentary structures in sand dunes, and internal layering on radar profiles due to density variations is commonly reported from polar ice sheets (MORSE et al. 1998). For these reasons, one should be cautious with interpreting the reflectors with respect to number and thickness of layers. This requires a detailed quantitative analysis of the electromagnetic signals received along the profiles.

Finally, a reflector seen on the radar profile does not necessarily represent a layer or boundary in the ground. A simple example is the direct waves, the first events on the trace, which represents the signal from transmitter antenna directly through air and along the ground surface to the receiver antenna. Signals from surface reflectors also generate reflections. Although the main part of the pulse energy is directed down into the ground, some electromagnetic waves will also propagate through air. If surface objects are close enough (determined by the electromagnetic velocity of air and the time window), the system may pick up reflections from these targets. On the rock glacier, large surface boulders, neighbouring rock glaciers and the backing cliff walls represent possible surface reflectors. The GPR profiles from Prins Karls Forland show many examples of such reflections. On both rock glaciers 7 and 9, the longitudinal profiles show slanting linear events that are picked up from the limit of the time window (as deep as the profiles are displayed) and then closer to the surface further up along the profile. On rock glacier 9, it can be followed from 50 m to the end of the profile at 155 m (*1* in Fig. 3A). Along this distance, its depth changes from 53 m to 20 m. Given the differences in ground and air electromagnetic velocity, this implies that the surface target should be respectively about 106 and 40 m away at these points. This suits very well with the location of a cliff wall on the upper part of the talus cone. This reflector is also visible on transverse profile 9-3 as an umbrella structure, and the distance to the same cliff wall matches here as well. A similar analysis can be carried out for some reflectors on rock glacier 7 (*3* on Fig. 2A).

Multiples may arise when a reflector is especially strong. Such events may be picked out by considering the slope of point reflector hyperbolas and depth versus slope for slanting linear events. A single multiple is picked up at double travel time and with double slope of the primary reflection.

Interpretation of GPR transverse profiles on Prins Karls Forland

In the following, we refer to reflectors as amplitudes traced on the GPR profile, while a layer represents the interpretation of a radar reflector.

We have earlier suggested (BERTHLING et al. 2000) that the zone of no reflectors found on most longitudinal and transverse profiles may represent a relatively homogeneous ice layer. Such an interpretation fits the pattern of DC-resistivity sounding curves on Prins Karls Forland (BERTHLING et al. 1998, 2000) as well as present knowledge from core drillings through rock glaciers (VONDER MÜHLL & HOLUB 1992, ARENSON et al. 2002). However, this zone is at least partly an effect of the applied gain function. The AGC gain typically creates low-amplitude zones below strong signals, as exemplified beneath reflector 3 in Fig. 3D. The AGC time window used in the processing was about 500 ns.

In the zone of scattered reflectors, reflectors found on the transverse profiles often have counterparts on the longitudinal profiles (Tables 1–4). Thus, these reflectors probably represent real layers. Their apparent depth of between 10 and 20 m place them well within the rock glacier body. BERTHLING et al. (2000) propose that such layers originate on the surface and represent episodic events of burial of ice supersaturated layers or snow layers by debris, so that these layers are incorporated into the permafrost. This is supported by the general development of these reflectors sub-parallel to the surface. However, such a view is challenged by those reflectors on the transverse profiles that have a slope opposite or very different from that of the surface slope. Such features can be an effect of inherited structures, caused by shifting geometry of the talus cone through time and possibly past coalescing of smaller talus cones/ developing rock glaciers.

VONDER MÜHLL et al. (2000) found two strong reflection horizons on a GPR longitudinal profile from Murtèl-Corvatsch rock glacier. These horizons corresponded to the positions of the shear horizon and the bedrock interface, known from core drilling (VONDER MÜHLL & HOLUB 1992) and borehole deformation measurements. In fact, shear zones have been found on all rock glaciers where borehole deformation measurements have been made (WAGNER 1992, HOELZLE et al. 1998, ARENSON et al. 2002). It is tempting, therefore, to adapt the interpretation of the lower zone of reflectors more or less well developed across the profiles to one of these options. However, as will be discussed below, this is not necessarily straight forward. In general, the shear zones so far found in rock glaciers are themselves very different, so that a common radar signature of such zones is most likely difficult to obtain.

Both rock glaciers 12 and 15 are probably too thick for the bedrock interface to be detected with 50 MHz antennas, at least in the frontal areas. The relatively horizontal reflectors developed across the front of rock glacier 15 (*3* and *4* in Fig. 5B) at an AHASL of about 40–46 m may well be a shear horizon. Their depth place them beneath the zone of dipping reflectors seen on the longitudinal profile and interpreted by BERTHLING et al. (2000) to be

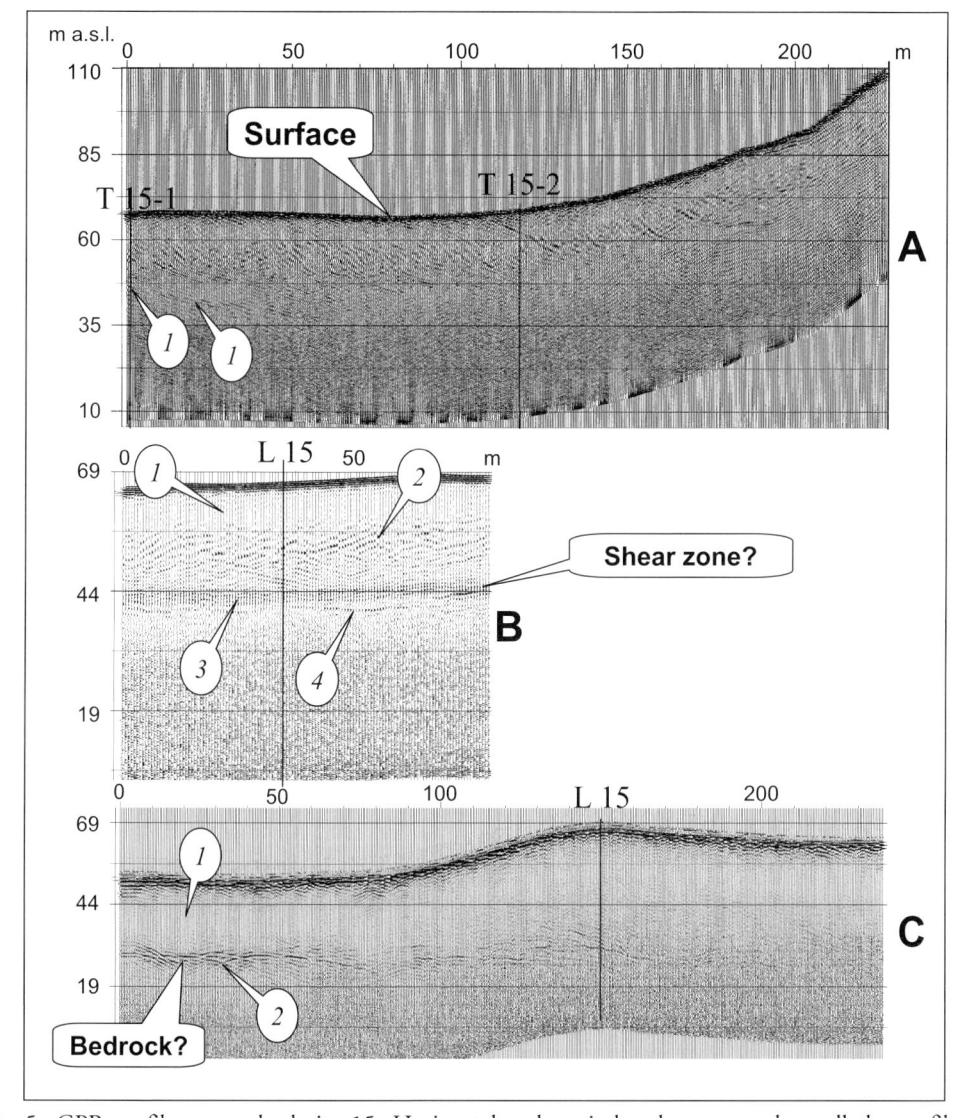

Fig. 5. GPR profiles on rock glacier 15. Horizontal and vertical scales are equal on all the profiles. Numbered pointers are referred to in Table 4 and in the text. **A**: longitudinal profile, **B**: frontal transverse profile, **C** transverse profile at rock glacier back. The appearance of this transverse profile is influenced by a shorter AGC window than what was used on the other profiles.

influenced by the dynamics of the rock glacier. The matching reflectors on the longitudinal profile have no obvious longitudinal development, although there is a vague series of reflectors that dip towards the surface and front of the rock glacier (*1* in Fig. 5A). Similarly, there are some vague reflectors at about the same depth (AHASL 33–45 m) on the front of rock glacier 12 (*3* in Fig. 4A). On rock glacier 15, the reflectors at an AHASL of 25–30 m (*2* in Fig. 5C) are probably bedrock, as the distance to the backing cliff here is small and its slope is not so steep.

The longitudinal profile of rock glacier 7 displays a continuous zone of reflectors from the front and 65 m inwards (*1* in Fig. 2A). While the rock glacier surface elevation rises about 10 m along this distance, the reflection horizon rises 3–4 m. The AHASL of this zone is about 25 m and the apparent depth is 18 m at the front. The AHASL of the zone is not compatible with a bedrock interface interpretation. On the other hand, its development is far more uneven than the shear horizon found by VONDER MÜHLL et al. (2000). The zone is not visible on the frontal transverse profile at all, making it even harder to interpret.

The best examples of profiles with partly continuous, deep reflectors are found on rock glacier 9. For these reflectors, a good match between longitudinal and transverse profiles can be demonstrated. We interpret these reflectors as the bedrock interface, and the smaller size of rock glacier 9 explains why bedrock is better visible here than on any of the other rock glaciers investigated. The interpretation is, however, open to some doubt. The longitudinal profile is ambiguous in the frontal part (between *2* and *3* in Fig. 3A), but the results from the frontal transverse profile places the bedrock at an apparent depth of about 30 m (*4* in Fig. 3B). The AHASL of this reflector varies between 11 and 19 m, which is in good agreement with the altitude of the strandflat in front of this rock glacier and reflector *2* in Fig. 3A. It is possible that the strong profiles found nearer surface in the transverse profile (*3* in Fig. 3B) might be a shear zone. They correspond partly to the set of reflectors at *3* in Fig. 3A. However, both the number of reflectors and their uneven geometry cause problems with such an interpretation. In the back of this rock glacier, the apparent depth the proposed bedrock reflector is only about 20 m and the AHASL varies between 23 and 30 m. This height is maintained to the lower part of the talus cone, but a general rise, continuing further up the talus cone, can be seen on the longitudinal profile. One would perhaps anticipate a more even outline of the bedrock interface than these numbers suggest. Especially, the rise in height from the front to the back of the rock glacier is surprising, and in this area a shear zone interpretation of the reflector might also have been favourable. Our reason to dismiss this interpretation is the fact that the apparent depth of the reflector increases upwards and into the talus cone. Within the study area, an abrupt break of slope between strandflat and backwall is found at the sites where no rock glaciers have developed and this must be anticipated also beneath them. The outline of the bedrock interface found on the rock glacier 9 longitudinal profile is in accordance with such a model, although the profile is not long enough for the break in slope to be detected. The alternative interpretation, a shear zone developed nearly to the apex of the talus slope, seems unlikely.

Recommendations for GPR studies on rock glaciers

From the above discussion, it is clear that GPR can give detailed information about internal structures within rock glaciers. For investigations of internal structures on rock glaciers, 50 MHz or 25 MHz antennas should be chosen. Especially for deep reflectors, our results with at 50 MHz antenna are ambiguous. The choice of 25 MHz antennas would certainly have improved the results on the deeper part of the profiles, due to better energy transfer to the deeper sections and the general suppression of clutter. However, 25 MHz antennas represent a challenge with respect to portability. Our investigations were performed in summer, as winter access to the site is difficult and probably hazardous with respect to avalanche danger. Winter surveys are nevertheless probably ideal. Both a sufficient snow cover which makes direct coupling between antenna and ground possible and a frozen active layer cause lower energy loss. For matching of longitudinal and transverse profiles and accurate terrain correction, exact positioning and topographic information is required. Simultaneous profiling with a differential GPS receiver is highly recommended.

Conclusions

From our study, we draw the following conclusions:
- The ground penetrating radar technique enables mapping of internal structures in rock glaciers.
- We achieved a depth penetration of about 30 m, and were partly able to see the bedrock interface.
- Reflectors could often be matched between longitudinal and transverse profiles so that strike and dip of single layers can be determined. Even true 3D-measurements are possible.
- Some reflectors could probably be interpreted as a shear zone.
- Although a 50 MHz antenna was used in our study, 25 MHz antennas are probably a better choice if quantitative investigations of layer thicknesses are not in focus and the logistical challenge can be met.

Acknowledgements

This study is part of the so-called Forlandet project, initiated in 1996 and led by Professor Johan Ludvig Sollid. The fieldwork was carried out in 1998 in connection with the field course in Arctic Geomorphology (GG365) in Ny-Ålesund, arranged by the Department of Physical Geography, University of Oslo. Senior engineer Trond Eiken, Department of Physical Geography, University of Oslo participated in the collection of radar data along with the students Espen Gudevang (University of Oslo) and Claudine Naguel (ETH/Zürich). The thorough comments of Dr. Norbert Blindow and one unknown referee significantly improved the paper. The study was financially supported by the Faculty of Mathematics and Natural Sciences, University of Oslo, and by a scholarship from the Norwegian Research Council on behalf of the Norwegian Polar Committee. The authors wish to thank all persons and institutions mentioned.

References

Annan, A. P. (1998): Ground penetrating Radar Workshop Notes. – Sensors and Software.
Annan, A. P. & Cosway, S. W. (1992): Ground penetrating radar survey design. – Proc. Symp. Applic. Geophys. Engin. Environm. Probl., SAGEEP'92, April 26–29, 1992, Oakbrook, IL: 329–251.
Arenson, L., Hoelzle, M. & Springman, S. (2002): Borehole deformation measurements and internal structure of some rock glaciers in Switzerland. – Permafrost Periglac. Process. **13**: 117–135.
Barsch, D. (1996): Rockglaciers. – 331 pp., Springer, Berlin.
Berthling, I., Etzelmüller, B., Eiken, T. & Sollid, J. L. (1998): Rock glaciers on Prins Karls Forland, Svalbard. I: Internal structure, flow velocity and morphology. – Permafrost Periglac. Process. **9**: 135–145.
Berthling, I., Etzelmüller, B., Isaksen, K. & Sollid, J. L. (2000): Rock glaciers on Prins Karls Forland, Svalbard. II: GPR soundings and the development of internal structures. – Permafrost Periglac. Process. **11**: 357–369.
Bristow, C. S., Bailey, S. D. & Lancaster, N. (2000): The sedimentary structure of linear sand dunes. – Nature **406** (6791): 56–59.
Daniels, D. J., Gunton, D. J. & Scott, H. F. (1988): Introduction to subsurface radar. – IEE Proc. **135** (5): 531–551.
Degenhardt, J. J., Giardino, J. R., Berthling, I., Isaksen, K., Ødegård, R. & Sollid, J. L. (2002): The internal structure of rock glaciers and geomorphologic interpretations: Yankee Boy Basin, Co, USA and Hiorthfjellet and Prins Karls Forland, Svalbard. – Geol. Soc. Amer. **34**, No. 6, Abstr. with Programs.
Haeberli, W., Wächter, H. P., Schmid, W. & Sidler, C. (1982): Erste Erfahrungen mit dem US Geological Survey – Monopuls Radioecholot im Firn, Eis und Permafrost der Schweizer Alpen. – VAW/ETH Zürich, Arbeitsh. **6**.
Haeberli, W., Hoelzle, M., Kääb, A., Keller, F., Vonder Mühll, D. & Wagner, S. (1998): Ten years after drilling through the permafrost of the active rock glacier Murtel, eastern Swiss Alps: Answered questions and new perspectives. – In: Lewkowicz, A. G. & Allard, M. (Eds.): Seventh International Conference on Permafrost. – pp. 403–409, Centre d'études nordiques, Université Laval, Yellowknife, Canada.
Haeberli, W. & Vonder Mühll, D. (1996): On the characteristics and possible origin of ice in rock glacier permafrost. – Z. Geomorph. N. F., Suppl.-Bd. **104**: 43–57.
Hoelzle, M., Wagner, S., Kääb, A. & Vonder Mühll, D. (1998): Surface movement and internal deformation of ice-rock mixtures within rock glaciers at Pontresina-Schafberg, Upper Engadin, Switzerland. – In: Lewkowicz, A. G. & Allard, M. (Eds.): Seventh International Conference on Permafrost. – pp. 465–471, Centre d'études nordiques, Université Laval, Yellowknife.
Isaksen, K., Ødegård, R. S., Eiken, T. & Sollid, J. L. (2000a): Composition, flow and development of two tongue-shaped rock glaciers in the permafrost of Svalbard. – Permafrost Periglac. Process. **11**: 241–257.
Isaksen, K., Mühll, D. V., Gubler, H., Kohl, T. & Sollid, J. L. (2000b): Ground surface-temperature reconstruction based on data from a deep borehole in permafrost at Janssonhaugen, Svalbard. – Ann. Glaciol. **31**: 287–294.
King, L., Fisch, W., Haeberli, W. & Wächter, H. P. (1987): Comparison of resistivity and radio-echo soundings on rock glacier permafrost. – Z. Gletscherkde. Glazialgeol. **23**: 77–97.
Lehmann, F., Vonder Mühll, D., van der Veen, M., Wild, P. & Green, A. (1998): True topographic 2-D migration of georadar data. – In: Proceedings of the Symposium on the Application of Geophysics to Environmental and Engineering Problems (SAGEEP). – pp. 107–114, Chicago.
Liestøl, O. (1976): Pingos, springs, and permafrost in Spitsbergen. – Norsk Polarinst. Årbok **1975**: 7–29.
Morse, D. L., Waddington, E. D. & Steig, E. J. (1998): Ice age storm trajectories inferred from radar stratigraphy at Taylor Dome, Antarctica. – Geophys. Res. Lett. **25**: 3383–3386.

TRABANT, P. K. (1984): Applied high-resolution geophysical methods. – 265 pp., Internat. Human Resourc. Developm. Corp., Boston.

VONDER MÜHLL, D. (1993): Geophysikalische Untersuchungen im Permafrost des Oberengadins. – VAW/ETH Zürich, Mitt. **122**, 222 pp.

VONDER MÜHLL, D. & HOLUB, P. (1992): Borehole logging in Alpine permafrost, Upper Engadin, Swiss Alps. – Permafrost Periglac. Process. **3**: 125–132.

VONDER MÜHLL, D. & HUGGENBERGER, P. (1997): Georadarmessungen auf dem kriechenden Permafrost des Blockgletschers Murtèl-Corvatsch. – VAW/ETH Zürich, Arbeitsh. **19**: 50–52.

VONDER MÜHLL, D., HAUCK, C. & LEHMANN, F. (2000): Verification of geophysical models in Alpine permafrost by borehole information. – Ann. Glaciol. **31**: 300–306.

WAGNER, S. (1992): Creep of Alpine permafrost, investigated on the Murtel Rock Glacier. – Permafrost Periglac. Process. **3**: 157–162.

Addresses of the authors: Dr. Ivar Berthling, Norwegian Water Resources and Energy Directorate, PO Box 5091 Majorstua, N-0301 Oslo, Norway. Dr. Bernd Etzelmüller, Associate Professor, Dept. of Physical Geography, University of Oslo, PO Box 1042 Blindern, N-0316 Oslo, Norway. Professor emeritus Johan Ludvig Sollid, Dept. of Physical Geography, University of Oslo, PO Box 1042. Cand. scient. Morgan Wåle, GeoPhysix, PO Box 56, N-1381 Vettre, Norway.

The radiomagnetotelluric method and its potential application in geomorphology

Andreas Hördt, Bonn and Gerhard Zacher, Köln

with 10 figures

Summary. The radiomagnetotelluric (RMT) method uses electromagnetic fields from existing radio transmitters to determine the electrical resistivity of the subsurface. Advanced inversion algorithms are available, providing high-resolution 2-D images. Three case histories illustrate the possible range of applications. An example from an industrial waste pit shows how RMT can delineate the boundaries of a waste deposit. During a groundwater exploration survey in Sweden, the combined interpretation of RMT and high-resolution reflection seismic data was used to determine the thickness of an aquifer. The third case history is a feasibility study on the applicability of RMT to find small cavities that may cause damage to agricultural vehicles. RMT might be particularly useful for geomorphologic problems in areas which are not easily accessible, because the RMT instrument is light compared to multichannel DC equipment. However, for rough terrain with high resistivities, such as rock glaciers or permafrost areas, further research is required before a routine application is possible. The high resistivities increase the contact resistance of the electrodes, cause a loss of spatial resolution and might require displacement currents to be considered during the interpretation. Topography can distort the data, and suitable correction algorithms have to be developed or adapted. Still, RMT might be a promising alternative in many situations, due to the dramatic logistical simplification compared to DC measurements.

Zusammenfassung. Die Radiomagnetotellurik (RMT) Methode verwendet elektromagnetische Felder vorhandener Radiosender, um den spezifischen elektrischen Widerstand des Untergrundes zu bestimmen. Für die Auswertung stehen Inversionsprogramme zur Verfügung, welche hochauflösende Abbilder des Untergrundes liefern. Drei Fallstudien veranschaulichen die möglichen Anwendungsgebiete. Ein Beispiel von einer industriellen Mülldeponie zeigt, wie RMT die lateralen Grenzen einer Ablagerung kartieren kann. Bei einer Messung zur Grundwasserexploration in Schweden konnte durch gemeinsame Interpretation mit hochauflösender Reflexionsseismik die Mächtigkeit eines Aquifers bestimmt werden. Beim dritten Beispiel handelt es sich um eine Machbarkeitsstudie, in der die Anwendbarkeit der RMT zur Detektion von Hohlräumen, welche Schäden an Landfahrzeugen verursachen können, untersucht wird. Die RMT-Methode könnte besonders in schwer zugänglichen Gebieten für geomorphologische Fragestellungen attraktiv sein, weil das Instrument verglichen mit einer Multikanal-Geoelektrik-Apparatur relativ leicht ist. Bevor die Methode jedoch routinemäßig in Gebieten mit starker Topographie oder hohen elektrischen Widerständen, wie z.B. Blockgletscher oder Permafrost, angewandt werden kann, sind weitere Forschungsarbeiten notwendig. Hohe Widerstände verschlechtern die elektrische Ankopplung der Elektroden an den Untergrund, verringern die räumliche Auflösung und machen die Berücksichtigung von Verschiebungsströmen erforderlich. Starke Topografie kann die Daten verzerren, so dass geeignete Korrekturen entwickelt oder adaptiert werden müssten. Wegen der logistischen Vereinfachung gegenüber der Gleichstromgeoelektrik könnte die RMT-Methode dennoch eine viel versprechende Alternative sein.

Introduction

Geophysical methods have been applied for many years in geomorphology, but mainly to permafrost-related problems. FISCH et al. (1977), later HAEBERLI & PATZELT (1982), KING et al. (1987) and BARSCH & KING (1989) used DC resistivity soundings and refraction seismics to show that rockglaciers usually do not have a pure ice core and thus have to be considered as periglacial phenomena. The conclusion was based on the high resistivity contrast between extremely resistive pure ice and frozen debris. Geophysical methods were also used to detect and to map permafrost as well as to determine thickness and even ice content of permafrost bodies (e.g. HAEBERLI & PATZELT 1982, KING 1984). More recently, high resolution refraction seismics and multichannel DC resistivity measurements with two-dimensional inversion algorithms have been applied, with a focus on permafrost (e.g. HAUCK 2001).

Several potential applications for geophysical methods exist that have not yet been fully exploited. These include the investigation of landslide bodies, karstic landforms and colluvia. Recently, geophysical methods have been used to investigate the structure and thickness of debris and complex valley fills (SASS & WOLLNY 2001, SCHROTT & ADAMS 2002, SCHROTT et al. 2003, HOFFMANN & SCHROTT 2002).

Most of these areas are not easily accessible with vehicles. Unless there is funding for a helicopter, the equipment has to be carried in difficult terrain. The seismic and multichannel DC resistivity equipment are portable, but the typical weight is 50–70 kg, which requires at least 3 people to carry. The radiomagnetotellurics (RMT) method might be a suitable alternative. The instrument is much lighter than seismics or DC equipment (about 15 kg), and can easily be transported by one person.

The RMT method determines the electrical resistivity (or its inverse, conductivity) of the ground, and thus the information is comparable to DC resistivity. It was originally developed for hydrogeological applications (e.g. MÜLLER 1983, THIERRIN & MÜLLER 1988). Later it was used for waste site investigation, engineering and archeological studies (ZACHER et al. 1996, HOLLIER-LAROUSSE et al. 1994, TEZKAN et al. 1996, TEZKAN 1999). The method has now been widely accepted as a powerful tool for near-surface investigation (PELLERIN & ALUMBAUGH 1997). Here, we explain the main principles of the method and review recent case histories to illustrate the variety of situations where RMT can be useful. We also discuss some of the particular difficulties that can arise when RMT is applied to geomorphology and indicate how they could be overcome.

The method

Radiomagnetotellurics uses electromagnetic fields from existing radio transmitters in the frequency range between 10 kHz and 300 kHz. The lower frequencies between 10 kHz and 30 kHz are called the very low frequency (VLF) range. The transmitters are normally used for navy communication, because the frequencies are sufficiently low to penetrate through sea water deep enough to be received by submarines. The power of these transmitters can be up to 1MW and, since the electromagnetic fields are transported through the cavity formed between the conductive ionosphere and the earth, they can be several thousands of km away from the actual survey location. They have been used many years in the well established VLF method, a review of which was given by MCNEILL & LABSON (1991). In the higher frequency range, long-wave broadcasting or time-signal antennas (SIEBEL 1984) are used.

The electromagnetic field of those vertical dipole antennas is polarised. The horizontal component of the electric field point towards the source, whereas the magnetic field only has a tangential component. To determine the electrical resistivity of the ground, the ratio between the two perpendicular components of the electric and magnetic field, the impedance, needs to be measured (Fig. 1). It depends on frequency and the electrical resistivity of the

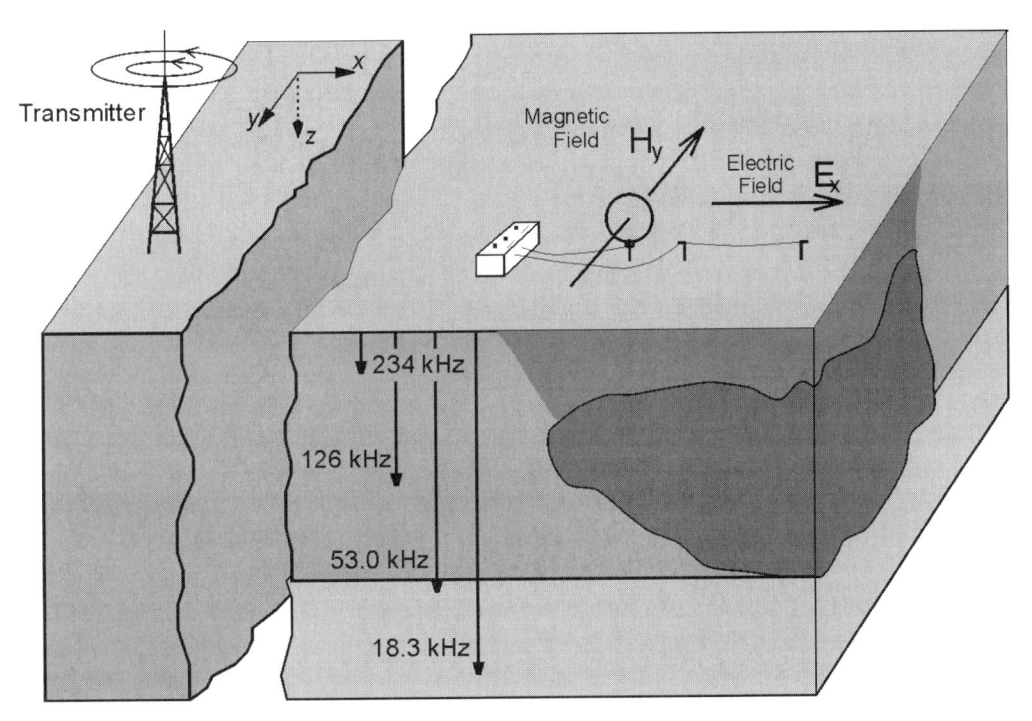

Fig. 1. Schematic sketch of the RMT method (after ZACHER et al. 1996). The electric field in the direction to the transmitter and the perpendicular magnetic field are measured for several frequencies. The arrows pointing into the ground illustrate the different penetration depths for four typical frequencies.

subsurface, but not on the amplitude of the transmitting field. Therefore, the impedance can be used to define an apparent resistivity with Cagniard's formula (CAGNIARD 1953) as:

$$\rho_{xy}(\omega) = \frac{1}{\omega\mu_0}\left|\frac{E_x}{H_y}\right|^2 = \frac{1}{\omega\mu_0}|Z_{xy}|^2 \qquad (1)$$

where ρ_{xy} denotes apparent resistivity, $\omega = 2\pi f$, where f is frequency, $\mu_0 = 4\pi 10^{-7}$ Vs/Am is the free-space magnetic permeability (Vs is Volt-second, Am is ampere-meter). E_x and H_y are the electric and magnetic field in the chosen coordinate system, and Z_{xy} is the impedance. The principle and the theory of radiomagnetotellurics are the same as for the magnetotelluric (MT) method, a review of which was given by VOZOFF (1991).

Over a homogeneous earth, the apparent resistivity does not depend on frequency and is equal to the true resistivity of the ground. If the subsurface is flat-layered, the apparent resistivity depends only on frequency, and not on location. Such a model is called a one-dimensional (1-D) model because the resistivity structure only depends on one coordinate. The highest frequency signals are attenuated most and thus do not penetrate deeply into the ground. They are representative of the uppermost layers. The lowest frequency signals penetrate deepest, and represent the deeper structure. By using several frequencies, a sounding curve is obtained. There is some similarity to DC resistivity soundings, where sounding curves are obtained by increasing the distance between transmitter and receiver electrodes.

The depth of penetration can be estimated from the skin depth, defined over a homogeneous halfspace by:

$$\delta = \sqrt{\frac{2\rho}{\mu_0\omega}} \approx 500\sqrt{\frac{\rho}{f}} \quad [m] \qquad (2)$$

The skin depth is the depth at which the amplitude of a harmonic electromagnetic wave is attenuated to 1/e of its value at the surface. For example, over a 100 Ωm halfspace at 10 kHz the skin depth is 50 m. The skin depth is not an exact measure of the maximum depth at which a layer or body might be detected, because the signal response at a given frequency also depends on deeper structures. However, the skin depth is still useful to obtain as a first estimate of the possible penetration.

Since the electric field lags behind the magnetic field, the impedance is a complex number, and a phase can be calculated in addition to the amplitude:

$$\varphi_{xy}(\omega) = \tan^{-1}\left(\frac{\text{Im}(Z_{xy})}{\text{Re}(Z_{xy})}\right) \qquad (3)$$

where 'Im' and 'Re' denote the imaginary and real parts of the complex impedance. Over a homogeneous halfspace, the phase is constant at 45°. Over inhomogeneous ground it varies with frequency. It approaches 45° in the high-and low frequency limits for theoretical models.

The measurements are usually carried out along profiles. A geologic strike direction typically is assumed, with the profiles laid out across strike. Two types of transmitter positions can then be distinguished: transmitters which are located in the prolongation of the profile produce an electric field parallel to the profile, and thus perpendicular to strike. The measured magnetic field from these transmitters is parallel to strike, and therefore this mode of measurement is called transverse magnetic (TM)-mode. Transmitters which are located on a line perpendicular to the profile produce an electric field parallel to strike. This mode is called transverse electric (TE)-mode. In the formalism set up above, for an x-directed strike, the TE- and TM-mode would correspond to xy- and yx-directions, respectively.

In Europe it is usually possible to find transmitters over the desired frequency range in both directions. The two modes contain different information and have different sensitivity with respect to lateral variations in the subsurface. Therefore, it is recommended to use both modes and combine the data during the interpretation (ZONGE & HUGHES 1991, BERDICHEVSKY et al. 1998). However, if only one mode is available for logistical reasons, the information is still substantial and useful.

The standard interpretation of the data is based on 2-D inversion. A two-dimensional model, where resistivity is assumed to vary with depth and in profile direction, is determined automatically through an iterative minimization procedure. Some software packages are freely available which have been originally developed for MT measurements (SMITH & BOOKER 1991, MACKIE & RODI 1996). The assumption of a 2-D model bears the risk of erroneous interpretation, because in reality the resistivity will always vary in all three directions. The limitations of interpreting data from a 3-D subsurface with 2-D models have been discussed by several authors (e.g. WANNAMAKER et al. 1984, PARK et al. 1983, BERDICHEVSKY et al. 1998). These include suggestions to invert subsets of the data, such as only TM mode data, to reduce the effects of 3-D structure.

In a general 3-D case, the horizontal electric and magnetic fields are related through a tensor relationship as:

$$\underline{E} = \underline{\underline{Z}}\, \underline{H} \qquad (4)$$

where

$$\underline{\underline{Z}} = \begin{pmatrix} Z_{xx} & Z_{xy} \\ Z_{yx} & Z_{yy} \end{pmatrix} \qquad (5)$$

is called the impedance tensor. The diagonal elements, Z_{xx} and Z_{yy} are zero in 1-D or 2-D situations where the coordinate system is along or perpendicular to strike, but nonzero in 3-D situations. Only one type of available RMT-equipment mentioned below is able to measure Z_{xx} and Z_{yy}, because these components are usually significantly smaller than the off-diagonal elements.

Three-dimensional inversion algorithms are currently being developed and tested, but are not yet applicable in routine work. Recently, NEWMAN et al. (2003), have successfully carried out a 3-D inversion of RMT data from an industrial waste deposit in Cologne. They had to ignore the diagonal elements of the impedance tensor because the equipment they used was unable to measure them.

Currently, only a few types of RMT equipment exist. The best established is the instrument developed by MÜLLER (1983) at the University of Neuchatel. The older version, which

Fig. 2. The Neuchatel equipment in the field. The geophysicist carries headphones for the manual-acoustic signal compensation. The picture is from a high resolution survey, with 1 m electrode distance. The electric field is measured perpendicular to the profile line, indicated by the measuring tape.

was used for most of the work cited above, uses analogue electronics and a manual-acoustic compensation to determine the apparent resistivity and phase at a given frequency (Fig. 2). A new version developed recently acquires the data automatically and includes an optional data logger. The equipment is not produced on a commercial basis, but might be available on request.

Another system called EnviroMT has been developed within an EU project by Metronix/Germany in cooperation with the University of Uppsala (PEDERSEN et al. 2003, BASTANI 2001). The instrument scans the entire frequency range of interest and automatically measures the full impedance tensor. In addition, a controlled source was developed to transmit signals of reasonable strength at frequencies between 1 kHz and 100 kHz, expanding the frequency range to lower frequencies where no radio transmitters exist. When a controlled source is used, the plane wave theory used to interpret the RMT data might not be valid if the distance between transmitter and receiver is too small (Zonge & Hughes 1991). Estimates of the required distances normally vary between 3 and 7 skin depths (Goldstein & Strangway 1975, Zacher et al. 1996) and may become as large as 20 skin depths for conductive layers over highly resistive units (Wannamaker 1997). Only one operational prototype of the the EnivoMT system seems to exist at the moment; a series production might be started if there is sufficient interest.

Other attempts have been made and are still ongoing to develop RMT instruments (e.g. Radic & Rath 1994), but no widespread commercial products exist at the moment. Although there are a number of convincing case histories, it seems that the method has not received the attention it deserves. The prototype equipment and inversion algorithms are probably unfamiliar to many potential users and, thus, perceived to be difficult to apply.

Case histories

Industrial waste site

The Mellendorf site is a former sandpit that has been filled with industrial waste during the late nineteensixties through to the mid-nineteeneighties. In 1987, it was found that the site never had been sealed at the bottom, which was 13 m below surface. It was decided to protect the site from rain water with a surface sealing of about 1.5 m thickness, consisting of soil, gravel, clay, and a synthetic fabric. This way, the industrial waste, mainly consisting of magnesium drosses, should be inhibited from leaking deeper into the ground. A natural seal in the form of a clay layer exists at a depth of about 60 m; above there is an inhomogeneous mixture of sand and gravel. The RMT survey was carried out in 1996 together with transient electromagnetic (TEM) measurements. The idea was to combine the deeper penetrating TEM data with the shallower RMT data to investigate whether and to what extent waste had penetrated below the assumed bottom. The full details of the RMT interpretation were given in Tezkan et al. (2000) and the joint interpretation with the TEM data has been described by Hördt et al. (2000).

Fig. 3 shows a 2-D inversion result obtained with the program by Mackie & Rodi (1996). The algorithm does not produce sharp boundaries, because it uses an overparameterized model with a controlled roughness. The profile crosses the deposit approximately in the middle. The edges of the deposit are known to be at 0 m and 180 m on the figure. The deposit appears as a conductive zone, embedded in a more resistive environment. Fig. 4 shows the fit of some of the data for the model of Fig. 3. The data are represented as apparent resistivities and phases along the profile. TM mode data of the highest and second lowest frequencies are shown; the full data set consists of four frequencies for each polarisation (20.3, 57.7, 128.0 and 207 kHz for N-S transmitters, 16.0, 60.0, 126.8 and 177.0 kHz for EW-transmitters). The fit is reasonable and typical for such a data set. Attempting to explain more detail in the measured data would risk an overinterpretation, because the data contain noise and could be influenced by 3-D structure, which is not considered in the inversion.

2D-JOINT-INVERSION Profil 0

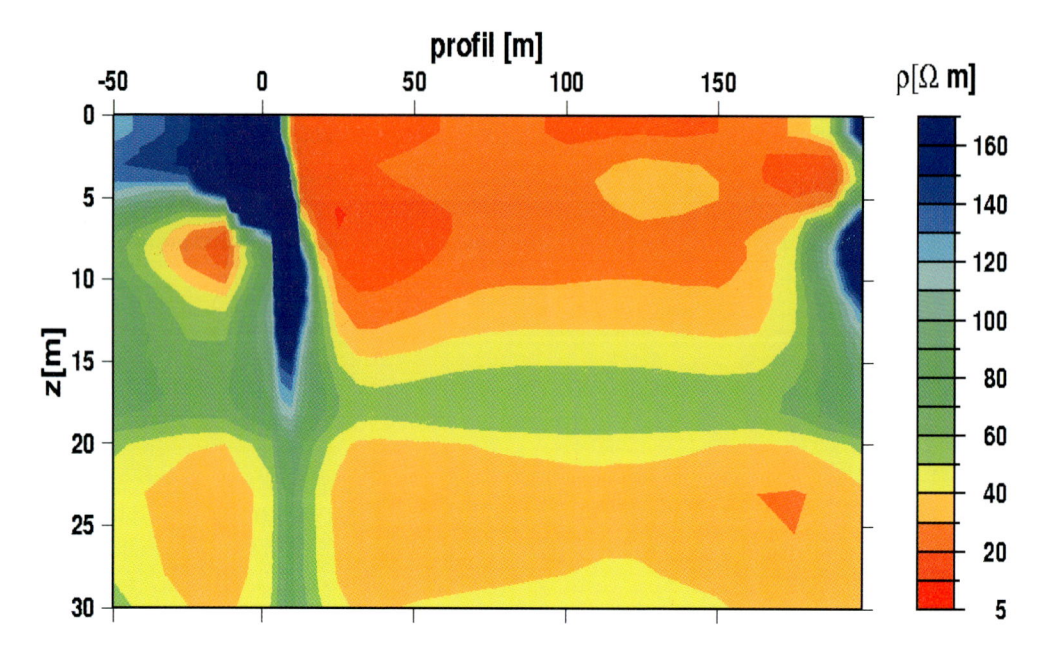

Fig. 3. 2-D inversion result of both TE and TM mode for a profile crossing the Mellendorf industrial waste site (after HÖRDT et al. 2000). The result was obtained with the code by MACKIE & RODI (1996).

At 0 m, the transition from the background into the waste may be disturbed by artifacts caused by a road and a corresponding distortion of the natural environment. A metallic culvert or a buried phone line is unlikely, because the signature of such distortions is different from the behaviour of the data shown in Fig. 4 (RECHER 1998). The transition looks similar to the undisturbed eastern edge at 180 m, indicating that potential artifacts are not severe. The conductive feature at − 20 m in 10 m depth was attributed to a clay lens, which are known to exist in this area. Apparently, the bottom of the waste at approx. 15 m (13 m plus the seal at the surface) is also visible. However, there is another conductive zone below 20 m. The resolution of the RMT method at this depth is not sufficient to determine whether these two conductive zones are really separated, or whether there is in fact one large zone. The resolution in this case is determined by the penetration depth of the lowest frequency (which penetrates deepest) rather than by the number of frequencies. The skin depth assuming an average resistivity of 10 Ωm, and taking the lowest frequency of 16 kHz, is 12,5 m, much less than the zone of interest. Forward modelling is clearly required to decide on the existence of the deeper conductor.

As mentioned above, TEM measurements were carried out to obtain a better resolution of the depth range below 15 m. These data were modelled jointly with the RMT data using a trial-and error procedure. Since the TEM data are sensitive to the northern and southern

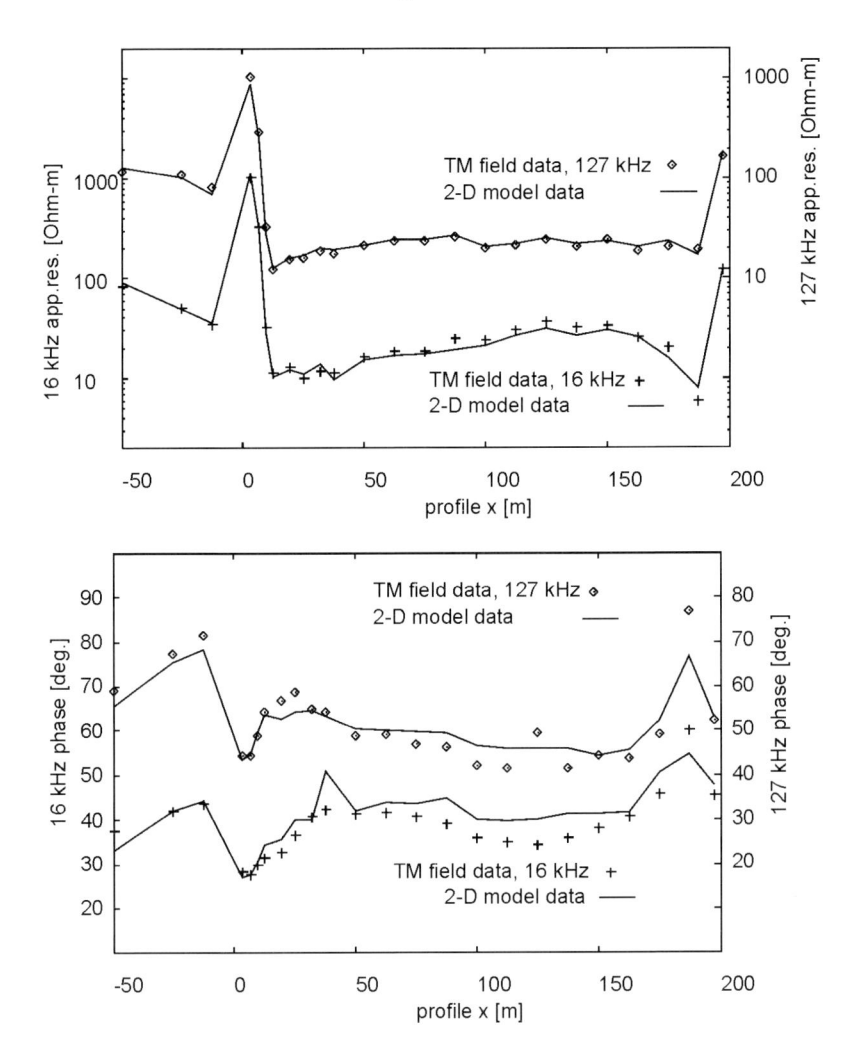

Fig. 4. Measured (symbols) and calculated (lines) data for the model of Fig. 3 (HÖRDT et al. 2000). Displayed are TM mode data for 2 of the 4 frequencies. Note that the scale is shifted between the two frequencies, as indicated at the left and right edges of the panels.

edges of the waste body, a 3-D model was constructed that includes those edges. There are no variations in the waste body itself in the direction perpendicular to the section. The numerical simulations were carried out using the 3-D finite-difference code by DRUSKIN & KNIZHNERMAN (1988). The RMT data were simulated with the same 3-D model using the code by MADDEN & MACKIE (1989). Although the RMT data are not influenced by the off-plane edges, a 3-D model was chosen in order to obtain one consistent model for both data sets.

A vertical section through the resulting model, which explains both data sets reasonably well, is shown in Fig. 5a. The resistivity structure is now described by blocks of constant resistivity rather than by a smooth distribution, because this is more practical in a trial-end error procedure. Currently, no automatic joint inversion algorithm for TEM and RMT data exists, and thus the models had to be constructed manually.

The new model now has a clear separation between an eastern and a western part of the deposit. The two separate conductive zones above and below 20 m which appeared in Fig. 4 are not necessary to fit the data. Apparently, the separation was an artifact produced by the inversion, which was run with a 10 Ωm halfspace starting model. The low resolution caused by a lack of lower frequencies allows different models in this depth range which yield an equally good data fit. However, conductive blocks below 15 m are required to fit the data. Additional calculations were carried out to test the sensitivity of the data with respect to these blocks, i.e. to see how the data fit changes if the blocks are moved. One example is the model in Fig. 5b, where the 5 Ωm block has been moved up by several metres.

The data fits corresponding to the two models in Fig. 5 are shown in Fig. 6. The difference between the two models is not dramatic anywhere, but in the lower frequency phases the fit by model 5a is superior to the one by model 5b. The difference is similar for the TEM data (not shown here). Combining the evidence of both data sets it was concluded that there is a significant amount of low-resistive material below the assumed bottom of the deposit. The

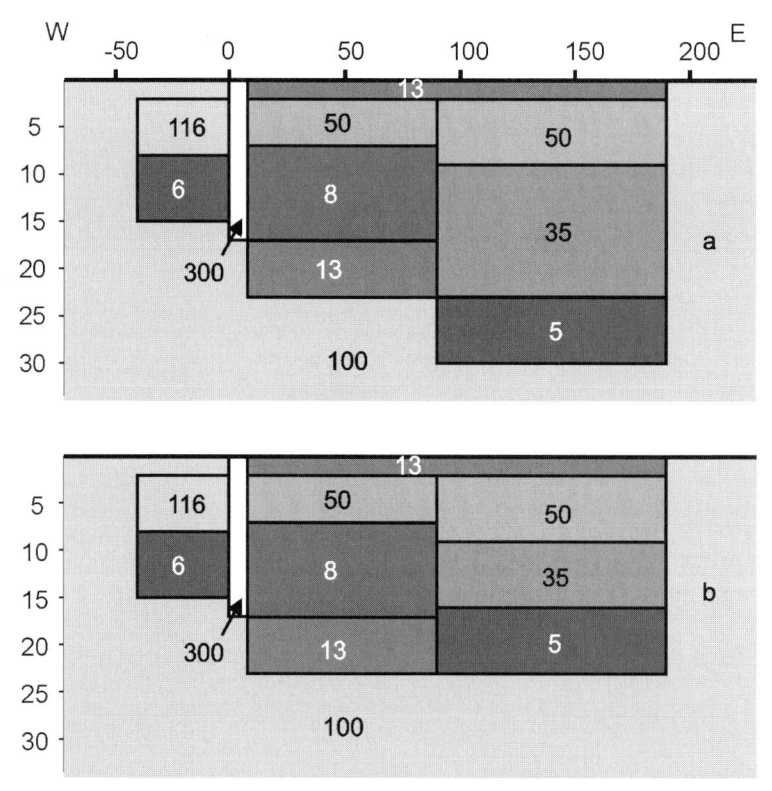

Fig. 5. Blocked resistivity models constructed by trial – and – error modeling to explain both RMT and TEM data sets (after HÖRDT et al. 2000). The top panel is the best-fit model. The lower model was used to test the influence of the conductive block on the calculated data. Numbers are resistivities in Ωm.

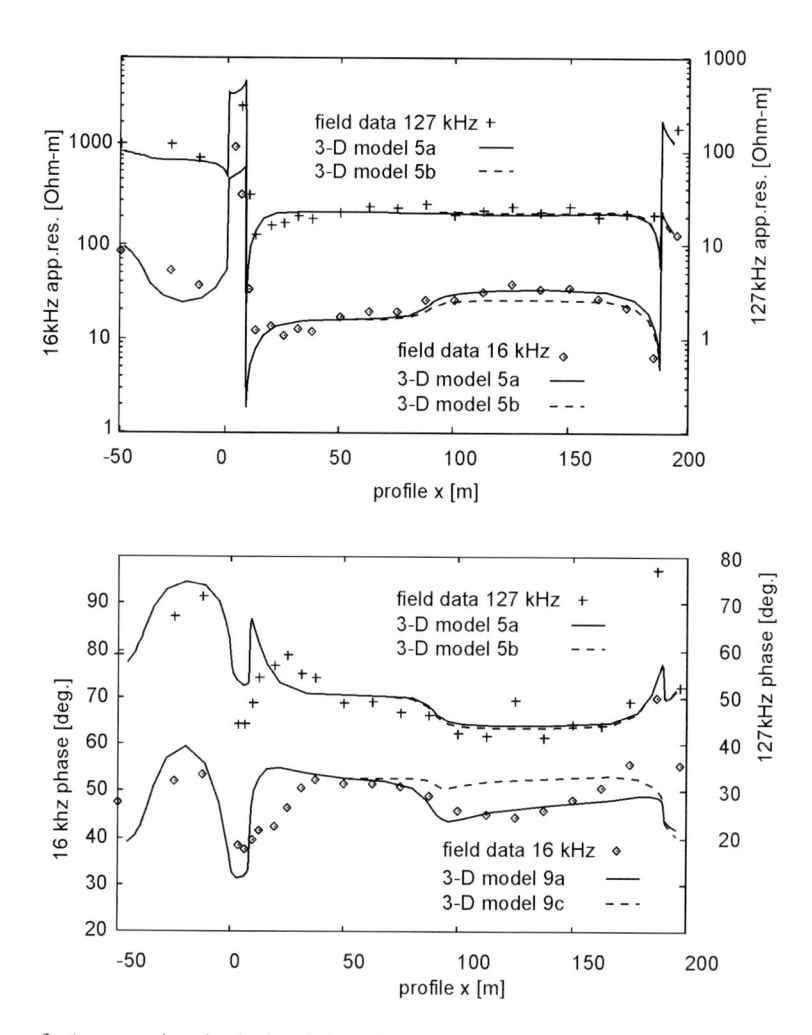

Fig. 6. Data fit (measured and calculated data) for the two resistivity models shown in Fig. 5.

decrease in resistivity is possibly caused by a leakage of waste into deeper formations. A geologic cause such as a clay lens might be an alternative explanation, although the spatial correlation with the waste deposit seems to favour a direct relationship.

Ground water exploration

A ground water exploration survey was carried out in the vicinity of Skediga/Sweden by the University of Uppsala. The measurements and details of the interpretation have been described by PEDERSEN et al. (2003), and BASTANI (2001). The aim of the survey was to determine the thickness of a sand/gravel aquifer, embedded between a thick clay layer on top and the crystalline basement below. Prior to the RMT measurements, a high-resolution seismics

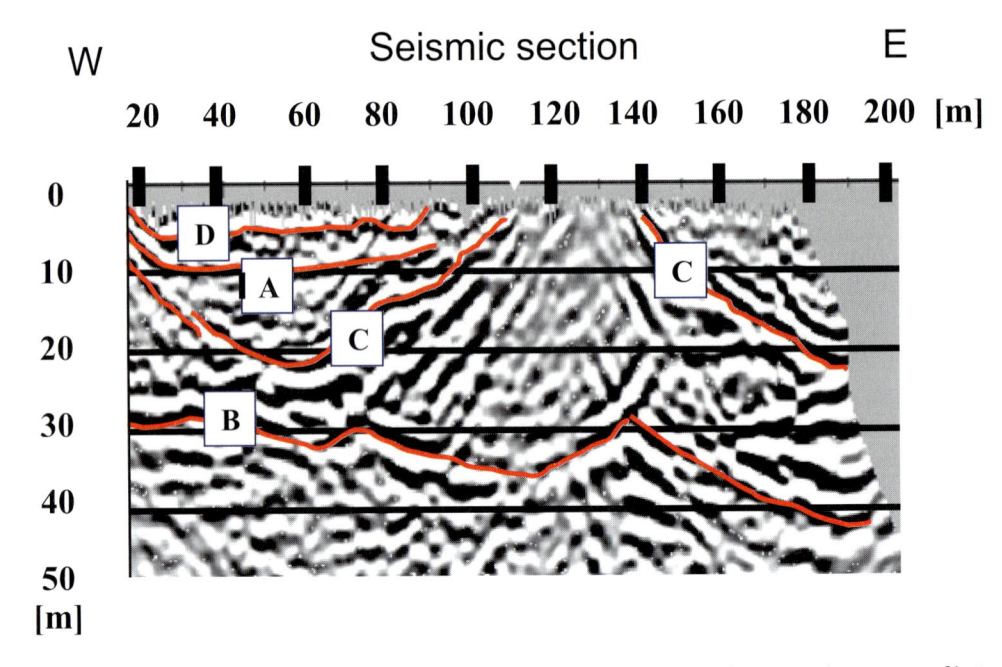

Fig. 7. Migrated high resolution reflection seismic section along the groundwater exploration profile in the Skediga area. Reflector B was interpreted as sand/gravel-crystalline basement interface. Reflectors C were taken to represent clay-sand/gravel interfaces, and reflectors A and D to represent thin sand layers within the clays (from PEDERSEN et al. 2003).

survey had been conducted. The survey and data processing was described in detail by JUHLIN et al. (2000). Fig. 7 shows the section along the same profile where the RMT measurements were taken, and the interpretation by JUHLIN et al. (2000). A lot of detail can be recognized, a fact that has been attributed by the authors to the use of small 50 g dynamite sources and a high spatial sampling of 1 m.

The important reflectors that have been identified are denoted C and B. Reflector C was interpreted as the transition from the clay layer to the sand aquifer, whereas reflector B was attributed to the crystalline basement. Reflectors A and D have been interpreted as thin sand layers within the clay, but these are not essential for the main conclusions.

The RMT data were acquired in two frequency ranges, using the existing VLF and radio transmitters in the high frequency range between 14 kHz and 250 kHz, and the controlled source for the low frequencies between 1 kHz and 14 kHz. It was found that the plane-wave assumption does not hold for the lowest frequencies. As mentioned above, it is useful if the source field can be considered a plane wave because fast and stable forward modelling and inversion codes exist. It is always difficult to determine the optimum distance to the source a priori, because it does not only depend on some average resistivity, but also on the particular layering. In this case, the source was located 350 m south of the profile, which apparently was too close for the lowest frequencies. Therefore, only frequencies above 4 kHz were used for inversion.

Another detail deviating from the Mellendorf case history is that determinant data were inverted instead of TE and TM mode. This means that apparent resistivities and phases were calculated from the determinant of the impedance tensor (eq. 5). The determinant represents an average of all current directions, is independent of the chosen coordinate system, and for 2-D models, the phases and logarithms of the apparent resistivities are just arithmetic means of the TE and TM phases and logarithmic apparent resistivities. Inversion of the determinant mode avoids the complication of having to fit different data sets with one model. The authors also claim from empirical observations that the determinant inversion provides higher resolution than the joint inversion of TE and TM mode.

Fig. 8 shows the inversion result, which was obtained with a modified version of the code by SIRIPUNVARAPORN & EGBERT (2000). Like the program used in the previous example, the algorithm uses an overparameterized model including a smoothness constraint, and thus the image does not yield sharp boundaries. However, the main seismic reflectors correspond well with the transitions between zones of different resistivities. The drillhole indicated at 105 m is in fact 50 m north of the profile. There, the basement depth is 35 m, which supports the interpretation that both reflector B and the 300 Ω-m contour correspond to the basement. There is a disagreement in the interpretation of reflector B towards the east. Whereas, in the seismic interpretation, the reflector continues at depths between 30 and 40 m, the resistivity image suggests an interpretation where the basement is 20 m shallower. In the middle of the profile, a comparison is difficult, because the seismic image is somewhat distorted and the RMT inversion might be affected by 3D effects. It looks as if there might be fault zone at 135 m, a possibility which has not been discussed by the original authors.

A penetration test at coordinate 200 met solid rock at 25 m depth. While this was originally interpreted as a boulder, the RMT interpretation now suggests that in fact the basement was hit, and that the seismic interpretation needs to be refined. When looking at the seismic section in Fig. 7 alone, it seems indeed unclear how to continue reflector B. The RMT image clearly facilitates the interpretation. The case history shows a remarkable amount of detail which can be extracted from RMT data, and can nicely be integrated with seismic data.

Cavity detection

This case history is a feasibility study rather than a routine application. Our aim was to assess whether the RMT method might be able to find shallow cavities. Such cavities had caused some damage when they collapsed under the weight of farmers' vehicles. Looking for cavities is a challenge for RMT, because inductive methods are generally less sensitive to resistive structures than to conductive features. Since RMT is a quick and relatively cheap method, it was considered worthwhile to make an attempt at finding such cavities.

In a first step, the apparent resistivities for all frequencies were displayed as a map (Fig. 9). Two main features are prominent: the separation into a northern and southern area with different resistivity ranges, and a fault in N-S direction approximately in the middle of the area known as 'Ophertener Sprung'. The cavities are expected to be aligned along the fault. Two-dimensional inversions were done for the short W-E profiles. Zones of anomalously

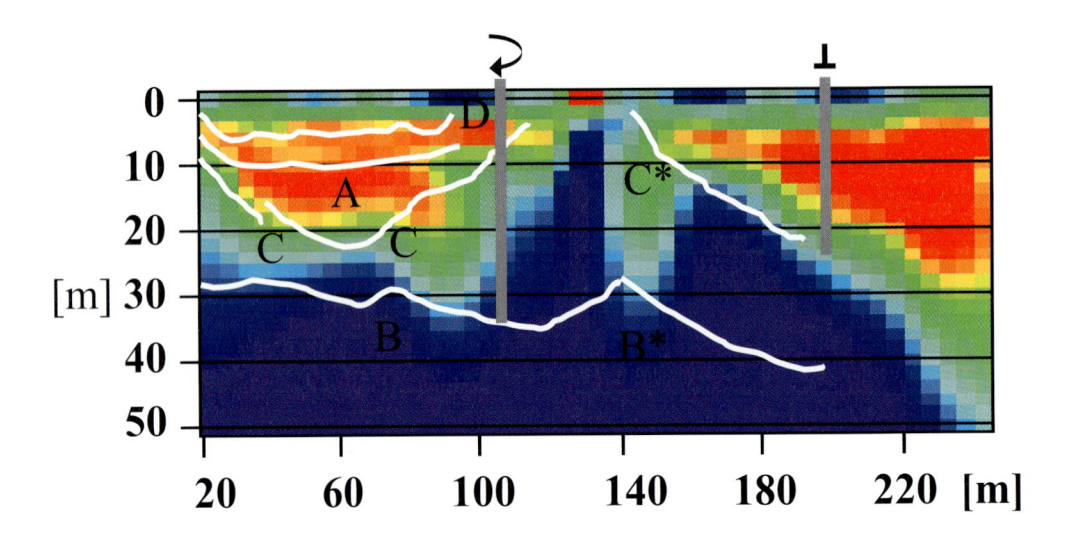

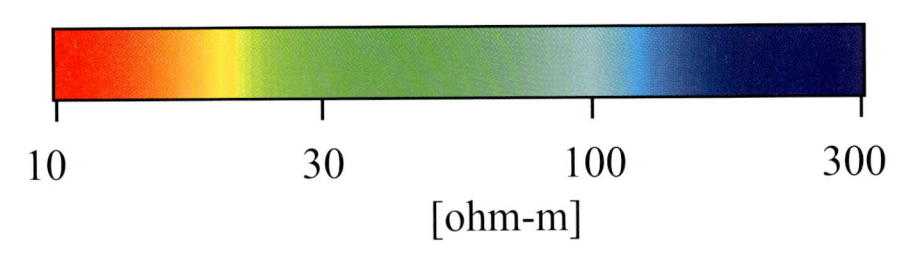

Fig. 8. 2-D inversion result of the RMT data along the same profile as in Fig. 7, with the main seismic reflectors superimposed The bore-hole in the middle is located ca 50 meters to the north of the profile, where it reaches basement at 35 meters depth. The bore-hole to the east reaches basement at 25 meters (from PEDERSEN et al. 2003).

high resistivities within the first 5 m were then identified and categorised, resulting in recommendations for further penetration tests.

Fig. 10 shows one example of such a 2-D inversion result. In this case, the code by SMITH & BOOKER (1991) was used, that has similar smoothing properties as the ones described above. The anomaly at 15 m along the profile has been categorised as potentially caused by a void space. The penetration tests indeed found a relatively large cavity here between 0,6 m and 1,8 m depth. The full evaluation on a statistical basis has not yet been completed. Some of the recommended points have been false alarms, some were partly successful (i.e. smaller cavities were found which would not make up a serious risk). This was a low-budget, semi-

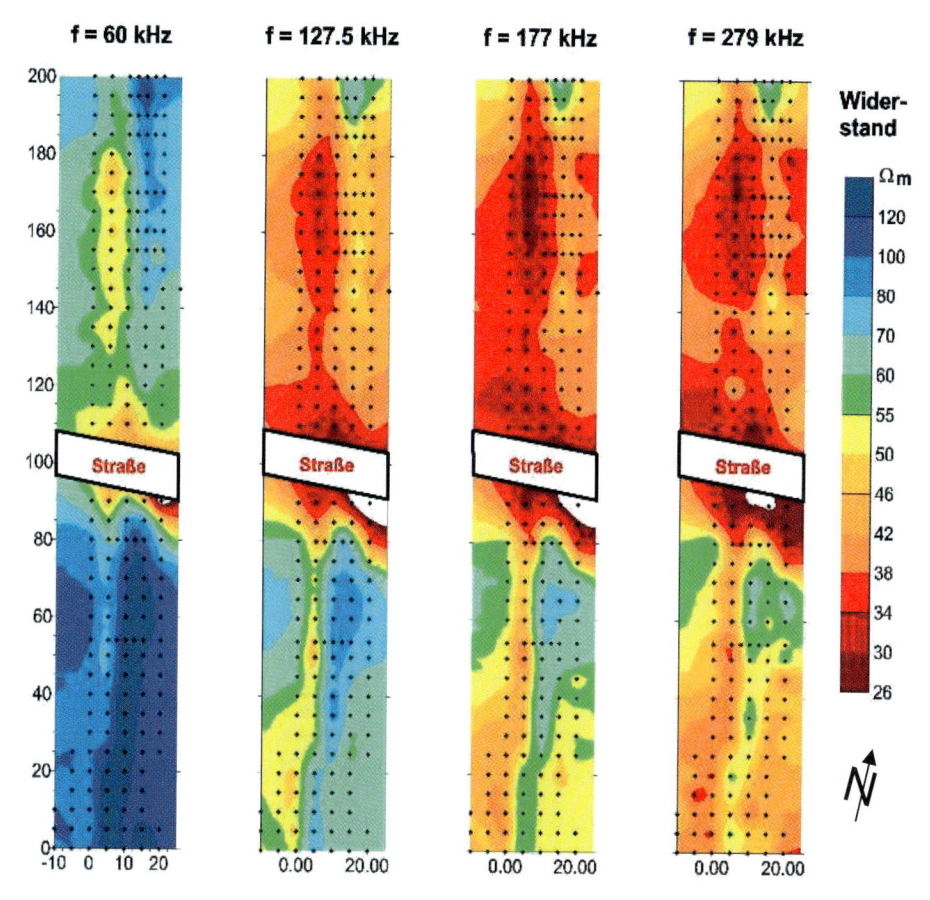

Fig. 9. Measured RMT apparent resistivity maps for four frequencies, colour coded over the survey area close to Bedburg near Cologne, Germany.

commercial experiment, without extensive modelling and accuracy evaluation. However, the results look promising, and since finding cavities is a problem of great practical relevance, the subject certainly deserves further attention.

Discussion

For the application of RMT in geomorphology, two aspects are of interest, which will be discussed below. First, RMT should be compared with the DC resistivity method, which is well established and gives comparable information. We emphasize some properties of the RMT method and point out in which situations RMT could be an alternative. Second, we discuss the influence of particular conditions typical for geomorphic problems, such as high resistivity and rugged terrain, on the applicability of RMT.

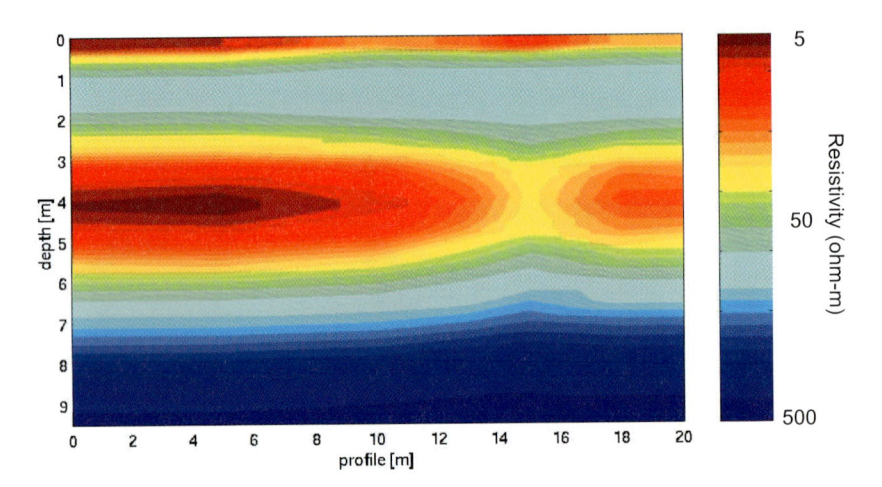

Fig. 10. 2-D inversion result for profile 200, corresponding to the topmost profile in Fig. 9.

Comparison with DC resistivity

Like radiomagnetotellurics, DC resistivity methods provide an image of the electrical resistivity of the subsurface. One difference between the methods is the physics. RMT is an inductive method, i.e. the currents are induced in the earth by the changing electromagnetic field. As a result, the currents can penetrate resistive layers easily, and thus, the method is not very sensitive to resistors. DC methods use a static current which does not easily penetrate resistive layers. Resistive layers may have a shielding effect, and do have a large influence on the measured voltage. The resolution properties in 2-D or 3-D situations are more complicated, but certainly different for DC and RMT.

A general advantage of the RMT method may be the possibility to establish whether the responses result from 1-D, 2-D or 3-D structure through analysis of the impedance tensor (REDDY et al. 1977). This information may be used to obtain more accurate models, for example by inverting only TM mode data in the case of a 3-D subsurface (WANNAMAKER et al. 1984).

The weight of the equipment and the data acquisition procedure are more practical aspects. Data acquisition speed is probably not a key issue. No systematic comparison is known, but 100–200 sounding points per day for RMT, and 200–300 electrode positions per day for multichannel DC, depending on equipment and conditions, are probably good estimates. The weight of the RMT instrument as shown in Fig. 2 is approx. 15 kg, including power supply and sensors, and can easily be carried by a single person. Multichannel DC equipment, depending on the number of electrodes and the length of the layout, can weigh 60–70 kg, and needs at least two persons to be carried. On flat terrain which is easily accessible with vehicles, the weight might not be important, but in difficult areas without roads, the difference in applicability may be huge.

The weight aspect might be decisive for application to geomorphologic problems. Rock glaciers, talus slopes and permafrost are often in a rugged mountain environment, difficult to access. It can be very expensive or impossible to carry out DC measurements, and the advantage of the lighter equipment will be substantial. However, those areas will also create additional difficulties for RMT, which will be discussed in the following section.

Resistive ground

One property that characterises many areas of interest for geomorphologists is the very high resistivity. This is particularly true in permafrost areas and rock glaciers, where resistivities between 10.000 Ωm up to several MΩm have been reported (e.g. FISCH et al. 1977, KING 1984, KING et al. 1987, HAUCK 2001). First experiments with the RMT method have been made by VONDER MÜHLL (1993) who used the equipment by MÜLLER (1983). He obtained encouraging results, but a final conclusion was not drawn because a systematic study was not possible and no inversion algorithm was available to him.

The main consequence of resistive ground is that the coupling of the electrodes can become difficult, which may be a problem for both DC and RMT methods. For the data acquisition system to give reliable and correct measurements, the contact resistance between electrodes and the ground should be much less than the input impedance of the analogue electronics, which is usually several MΩ. One possibility to decrease the contact resistance is to attach sponges soaked in salt water to the electrodes (e.g. HAUCK 2001). This requires some extra time, material and effort to devote to every electrode, which clearly is a disadvantage for any multielectrode system.

An alternative may be capacitive coupling. Alternating electric fields can be recorded with electrodes which do not have contact with the ground. The electrode forms a capacitor with the conductive earth. The impedance of a plate electrode does not depend on the ground resistivity any more, and is given by (SMYTHE 1968):

$$R = \frac{1}{i\omega C} = \frac{d}{i\omega \varepsilon \varepsilon_0 S} \qquad (6)$$

where ε_0 and ε are vacuum and relative electrical permittivity, S is the area of the plate and d the distance between the plate and the ground. Therefore, the contact resistance can be decreased by increasing the area of the electrode. A compromise between resistance and handling will be necessary, but in principle this approach is promising when the ground is very resistive (WAIT 1995). Capacitive electrodes have been considered for DC measurements, which are then carried out at low frequencies around several kHz (TABBAGH et al. 1993, SHIMA et al. 1995). At this frequency range, due to the small distance between source and receiver, inductive effects can be ignored and the recorded fields can be interpreted using DC theory. This does not hold for RMT measurements, where the source is considered infinitely far away, and the electromagnetic fields are purely inductive.

The use of capacitive electrodes for inductive methods has been studied by MACNAE & MCGOWAN (1991). The particular case for magnetotellurics around 12 Hz down to 0.1 Hz was discussed by MACNAE (1995), who concludes that portability of the electrodes will be the main limiting factor. For the higher frequencies used in RMT, the conditions should be much more favourable, and capacitive electrodes might become a suitable alternative to soaked sponges.

Another effect of high resistivities is the loss of resolution for inductive methods. The skin depth given in eq. (2) is also the wave length of the electromagnetic fields, and therefore some measure of spatial resolution. It increases with the square root of resistivity, and thus can become very large in rock glacier or permafrost environments. However, inductive methods have been used successfully in permafrost environments (e.g. HAUCK 2001), and a loss of resolution might well be compensated by the easier data acquisition.

A related observation is that resistivities measured in permafrost with inductive methods are often significantly lower than those measured with DC methods (MÜLLER 2002, pers. comm., HAUCK 2001). In horizontally layered regions, such phenomena can be explained by anisotropy, a higher resistivity in vertical than in horizontal direction due to fine layering. Since inductive methods are sensitive to horizontal resistivity only, they usually see lower values than DC methods, that respond equally to horizontal and vertical resistivity (JUPP & VOZOFF 1977). However, in permafrost regions a fine layering is unlikely, and the cause of the different resistivities is still unclear at this stage.

High resistivities can also affect an important assumption usually made in the interpretation of RMT data. The so-called displacement currents, that are caused by a time varying electrical field, are neglected. This assumption holds at sufficiently low frequencies and moderate resistivities. It fails for extremely high resistivities as encountered in permafrost regions. Recently, PERSSON & PEDERSEN (2002) found that ignoring displacement currents may lead to a bias in the interpretation if resistivities are larger than 10000 Ω-m. They also offered a solution by including the effects in a 1-D inversion routine. For 2-D and 3-D inversion, no codes including displacement currents are routinely available, and further research will be necessary to model the effects and possibly find corrections.

Rugged terrain

Finally, rugged terrain may cause distortions of the data. When measurements are done in a narrow valley, the mountains can act like a shield for the source signal, which can become very weak for transmitters in the direction perpendicular to the value. The second effect is that induction in the mountains cannot easily be modeled, because most of the standard inversion codes assume a planar surface. Topographic effects in low-frequency magnetotelluric data have been studied by WANNAMAKER et al. (1986), and a correction algorithm called 'stripping' has been developed by CHOUTEAU & BOUCHARD (1988). The algorithm determines a frequency-dependent distortion tensor based on forward modelling that includes topography and an assumed background model. However, it is not clear whether the correction can be transferred to the RMT case. The approach assumes that electromagnetic coupling between topography and the subsurface is small, which might not hold in the RMT frequency range.

Conclusions

The case histories have shown that the RMT method can be a powerful tool for near-surface investigation. Good quality data can be acquired at reasonable speed, and 2-D inversion codes are routinely available. Still, there are not many instruments in use, and only a few groups worldwide apply RMT routinely. One reason may be that RMT has to compete with DC resistivity measurements, which gives comparable information. For DC methods, multi-channel equipment and 2-D inversion are routinely available and easy to use. Therefore, people rely on the well established technique, where the physics appears less complicated than for RMT. However, RMT is not really difficult to apply, and comfortable 2-D inversion software exists as well.

Radiomagnetotellurics can be a valid alternative in difficult terrain which is not accessible by vehicles. Light equipment is available that can be carried and operated by a single person. Some research will be necessary to clarify the effect of resistive ground and rugged terrain. In essence, RMT could be a valuable additional tool for geomorphologists. For some applications, existing hardware and software can be used; others require further developments or theoretical insight. Exciting experiments are waiting, such as the application to permafrost or talus slopes, that could be fruitful for both geophysicists and geomorphologists.

Acknowledgements

Laust B. Pedersen provided some of the material presented here. We also would like to thank Frank Bosch and Imre Müller for useful discussions. The work was partly sponsored by the Deutsche Forschungsgemeinschaft (Project Ho1506/3-1, Gr1443/1-1).

References

Barsch, D. & King, L. (1989): Origin and geoelectrical resistivity of rockglaciers in semi-arid subtropical mountains (Andes of Mendoza, Argentina). – Z. Geomorph. N. F. **33**: 151–163.

Bastani, M. (2001): EnviroMT – a New Controlled Source Radio Magnetotelluric System. – Acta Universitatis Uppsaliensis **32**, 179 S.

Berdichevsky, M. N., Dmitriev, V. I. & Pozdnjakova, E.E. (1998): On two-dimensional interpretation of magnetotelluric soundings. – Geophys. J. Int. **133**: 585–606.

Cagniard, L. (1953): Basic theory of the magnetotelluric method of geophysical prospecting. – Geophysics **18**: 605–635.

Chouteau, M. & Bouchard, K. (1988): Two-dimensional terrain correction in magnetotelluric surveys. – Geophysics **53**: 854–862.

Druskin, V. L. & Knizhnerman, L. A. (1988): A spectral semi-discrete method for the numerical solution of 3D nonstationary problems in electrical prospecting. – Phys. Solid Earth **24**: 641–648.

Fisch, W. sen., Fisch, W. jun. & Haeberli, W. (1977): Electrical DC resistivity soundings with long profiles on rock glaciers and moraines in the alps of Switzerland. – Z. Gletscherkde. Glazialgeol. **13** (1/2): 239–260.

Goldstein, M. A. & Strangway, D.W. (1975): Audio-frequency magnetotellurics with a grounded electric dipole source. – Geophysics **40**: 669–683.

Haeberli, W. & Patzelt, G. (1982): Permafrostkartierung im Gebiet der Hochebenkar-Blockgletscher, Oberurgl, Ötztaler Alpen. – Z. Gletscherkde. Glazialgeol. **18**: 127–150.

Hauck, C. (2001): Geophysical methods for detecting permafrost in high mountains. –Mitt. Versuchsanst. Wasserbau, Hydrol. Glaziol., ETH Zürich **171**, 204 S.

Hoffmann, T. & Schrott, L. (2002): Modelling sediment thickness and rockwall retreat in an Alpine valley using 2D-seismic refraction (Reintal, Bavarian Alps). – Z. Geomorph. N. F., Suppl.-Bd. **127**: 153–173.

Hollier-Larousse, R., Lagabrieelle, R. & Levillain, J. P. (1994): Utilisation de la radio-magnetotellurique pour la reconnaissance en site aquatique. – J. Appl. Geophys. **31**: 72–84.

Hördt, A., Greinwald, S., Schaumann, S., Tezkan, B. & Hoheisel, A. (2000): Joint 3D interpretation of radiomagnetotelluric (RMT) and transient electromagnetic (TEM) data from an industrial waste deposit in Mellendorf, Germany. – Eur. J. Env. Eng. Geophys. **4**: 151–170.

Juhlin, C., Palm, H., Müllern, C.-F. & Wållberg, B. (2000): High resolution reflection seismics applied to detection of groundwater resources in glacial deposits, Sweden. – Geophys. Res. Lett. **27**: 1575–1578 .

Jupp, D. L. B. & Vozoff, K. (1977): Resolving anisotropy in layered media by joint inversion. – Geophys. Prosp. **25**: 460–470.

King, L. (1984): Permafrost in Skandinavien – Untersuchungsergebnisse aus Lappland, Jotunheimen und Dovre/Rondane. – Heidelberger Geogr. Arb. **76**, 174 S.

King, L., Fisch, W., Haeberli, W. & Waechter, H. P. (1987): Comparison of Resistivity and Radio-Echo soundings on rock glacier permafrost. – Z. Gletscherkde. Glazialgeol. **23**: 77–97.

Mackie, R. & Rodi, W. (1996): A nonlinear conjugate gradients algorithm for 2-D magnetotelluric inversion. – Eos, Transact., Amer. Geophys. Union **77**, Issue 46, Suppl., pp. 155.

Macnae, J. (1995): Design considerations for capacitive electrodes. – Workshop «Electrodes» held at the Centre De Recherches Geophysiques Garcy, April 24–29, Proceedings (unveröff.), 32–35.

Macnae, J. C. & McGowan, P. D. (1991): Quantitative resistance and capacitive electrodes: new developments in inductive source resistivity. – Explor. Geophys. **22**: 251–256.

Madden, T. R. & Mackie, R.L. (1989): Three dimensional magnetotelluric modeling and inversion. – Proc. IEEE **77** (2): 318–333.

McNeill, J. D. & Labson, V. F. (1991): Geological Mapping using VLF radio fields. – In: Nabighian, M. (Hrsg.): Electromagnetic Methods in Applied Geophysics, Vol.2: Application, Part B. – 521–640, Tulsa, OK.

Müller, I. (1983): Anisotropic properties of rocks detected with electro-magnetic VLF (Very Low Frequency). – Internat. Symp. Field Measurements in Geomechanics, Special Publ.: 273–282; Zürich.

Newman, G. A., Recher, S., Tezkan, B. & Neubauer, F. M. (2003): Threedimensional inversion of a scalar radio magnetotelluric field data set. – Geophysics, submitted.

Park, S. K., Orange, A. S. & Madden, T. R. (1983): Effects of three-dimensional structure on magnetotelluric sounding curves. – Geophysics **48**: 1402–1405.

Pedersen, L. B., Bastani, M., Dynesius, L. & Oskooi, B. (2003): Ground water exploration using the high resolution EnviroMT technique. – Geophysics, submitted.

Pellerin, L. & Alumbaugh, D. (1997): Tools for electromagnetic investigation of the shallow subsurface. – The Leading Edge **16**: 1631–1638.

Persson, L. & Pedersen, L. B. (2002): The importance of displacement currents in RMT measurements in high resistivity environments. – J. Appl. Geophys. **51**: 11–20.

Radic, T. & Rath, V. (1994): Radiomagnetiche Sondierungen (RMS) – ein neuartiges EM-Messkonzept für den oberflächennahen Untergrund. – 3. Seminar Umweltgeophysik, Neustadt/Weinstr. Deutsche Geophysikalische Gesellschaft, Potsdam.

Recher, S. (1998): Untersuchung der Anwendbarkeit der Radiomagnetotellurik-Methode auf die Lokalisierung von kontaminiertem Boden. – 112 S., Diplomarbeit (unveröff.), Universität zu Köln.

Reddy, I. K., Rankin, D. & Phillips, R. J. (1977): Three-dimensional modelling in magnetotelluric and magnetic variational sounding. – Geophys. J. Roy. Astr. Soc. **51**: 313–325.

Sass, O. & Wollny, K. (2001): Investigations regarding alpine talus slopes using ground-penetration radar (GPR) in the Bavarian Alps, Germany. – Earth Surf. Proc. Landforms **26**: 1071–1086.

Schrott, L. & Adams, T. (2002): Quantifying sediment storage and Holocene denudation in an alpine basin, Dolomites, Italy. – Z. Geomorph. N. F., Suppl.-Bd. **128**: 129–145.

Schrott, L., Hufschmidt, G., Hankammer, M., Hoffmann, T. & Dikau, R. (2003): Spatial distribution of sediment storage types and quantification of valley fill deposits in an Alpine basin, Reintal, Bavarian Alps, Germany. – Geomorphology (in press)

Shima, H., Texier, B., Kobayashi, T. & Hasegawa, N. (1995): Fast imaging of shallow resistivity structures using a multichannel capacitive electrode system: – 65th Ann. Internat. Mtg. Soc. Expl. Geophys., 377–380; Tulsa, OK.

Siebel, W. (1984): KW-Spezial-Frequenzliste. AM – SSB – CW – RTTY. See- und Flugfunk. – 256 S., Siebel Verlag, Wachtberg.

Siripunvaraporn, W. & Egbert, G. (2000): An efficient data-subspace inversion method for two-dimensional magnetotelluric data. – Geophysics **65**: 791–803.

Smith, J. T. & Booker, J. R. (1991): Rapid inversion of two and three dimensional magnetotelluric data. – J. Geophys. Res. **96**: 3905–3922.

Smythe, W. R. (1968): Static and dynamic electricity. – McGraw-Hill.

Tabbagh, A., Hesse, A. & Grard, R. (1993): Determination of electrical properties of the ground at shallow depth with an electrostatic quadropole: Field trials on archeological sites. – Geophys. Prosp. **41**: 579–598.

Tezkan, B., Goldman, M., Greinwald, S., Hördt, A., Müller, I., Neubauer, F. M. & Zacher, G. (1996): A joint application of radiomagnetotelluric and transient electromagnetics of a waste deposit in Cologne (Germany). – J. Appl. Geophys. **34**: 199–212.

Tezkan, B., Hördt, A. & Gobashy, M. (2000): Two dimensional inversion of radiomagnetotellurics data: selected case histories for waste site exploration. – J. Appl. Geophys. **44**: 237–256.

Tezkan, B. (1999): A review of environmental applications of quasi-stationary elctromagnetic techniques. – Surv. Geophys. **20**: 279–308.

Thierrin, J. & Müller, I. (1988): La méthode VLF/Resistiviée multifrequence, un example d'exploration hydrogéologique dans un synclinal crétacé a la Brevine. – Ann. Scient. l'Univ. Besançon, Mém. hors ser. **6**: 17–24.

Vonder Mühll, D. (1993): Geophysikalische Untersuchungen im Permafrost des Oberengadins. – Mitt. VAW-ETH Zürich **122**, 222 S.

Vozoff, K. (1991): The Magnetotelluric Method. – In: Nabighian, M. (Hrsg.): Electromagnetic Methods in Applied Geophysics, Vol. 2: Application, Part B. – pp. 641–711, Tulsa, OK.

Wait, J. R. (1995): Comments on: Benderitter et al., Application of the electrostatic quadripole to sounding in the hectometric depth range. – J. Appl. Geophys. **34**: 79–80.

Wannamaker P. E. (1997): Tensor CSAMT survey over the sulphur springs thermal area, Valles Caldera, New Mexico, USA, 2. Implications for CSAMT methodology. – Geophysics **62**: 466–476.

Wannamaker, P. E., Hohmann, G. W. & Ward, S. H. (1984): Magnetotelluric responses of three-dimensional bodies in layered earths. – Geophysics **49**: 1517–1533.

Wannamaker, P. E., Stodt, J. A. & Rijo, L. (1986): Two-dimensional topographic responses in magnetotellurics modeled using finite elements. – Geophysics **51**: 2131–2144.

Zacher, G., Tezkan, B., Neubauer, F. M., Hördt, A. & Müller, I. (1996): Radimagnetotellurics: a powerful tool for-waste site exploration. – Eur. J. Env. Eng. Geophys. **1**: 139–159.

Zonge, K. L. & Hughes, L. J. (1991): Controlled source audio-frequency magnetotellurics. – In: Nabighian, M. (Hrsg.): Electromagnetic Methods in Applied Geophysics, Vol. 2: Application, Part B. – pp. 713–809, Tulsa, OK.

Addresses of the authors: Andreas Hördt, Geologisches Institut der Universität, Bonn Fachrichtung Angewandte Geophysik, Nussallee 8, D-53115 Bonn, Germany. Gerhard Zacher, Umwelt- und Ingenieurgeophysik, Kellereiweg 20, D-51107 Köln, Germany.

Multi-method geophysical investigation of a sporadic permafrost occurrence

Christof Kneisel and Christian Hauck

with 5 figures

Summary. In the Bever Valley, Upper Engadine, eastern Swiss Alps, a multi-method geophysical investigation of a sporadic permafrost occurrence was performed during several field trips. The multi-method approach included the application of one-dimensional DC resistivity soundings, two-dimensional DC resistivity tomography, sledgehammer refraction seismics, electromagnetics and capacitive coupled geoelectrical profiling (OhmMapper). The results of the geophysical measurements confirmed permafrost occurrences below the timberline consisting of several thin permafrost lenses. The application of the different geophysical methods has proven to be a useful approach for characterising the sporadic permafrost distribution on a small scale, as each of the methods can be applied to investigate special aspects of the permafrost occurrence.

One-dimensional DC resistivity soundings are used to determine the overall vertical resistivity distribution in the area as a first indicator for the presence of permafrost. The combination of two-dimensional DC resistivity tomography and refraction seismic tomography yields highly reliable detection of subsurface permafrost occurrences on a small spatial scale.

Electromagnetic induction methods (such as the EM-31) may be used to map shallow permafrost occurrences over large areas provided that they are used in combination with single DC resistivity tomography measurements for calibration. On snow-covered and mainly flat terrain the OhmMapper may be used as alternative for mapping permafrost over large areas.

Zusammenfassung. Im Bevertal (Oberengadin, Schweizer Alpen) wurde ein sporadisches Permafrostvorkommen während mehrerer Feldkampagnen mit verschiedenen geophysikalischen Methoden untersucht (ein-dimensionale geoelektrische Sondierungen, zwei-dimensionale Widerstandstomographie, Hammerschlag-Refraktionsseismik, Elektromagnetik, geoelektrische Widerstandskartierung mit kapazitiver Ankopplung – OhmMapper). Die Ergebnisse der geophysikalischen Messungen bestätigen die Existenz eines sporadischen Permafrostvorkommens unterhalb der Waldgrenze, bestehend aus mehreren geringmächtigen Permfrostlinsen. Die Verwendung verschiedener geophysikalischer Methoden hat sich als sinnvoller Ansatz zur kleinräumigen Charakterisierung der Permafrostverbreitung erwiesen, da mit jeder Methode verschiedene Charakteristika des Permafrostvorkommens erfassbar sind.

Ein-dimensionale geoelektrische Sondierungen wurden angewendet um die vertikale Widerstandsverteilung als ein erster Hinweis für das Vorhandesein von Permafrost zu erfassen.

Die Kombination von zwei-dimensionaler Widerstandstomographie und Refraktionsseismiktomographie liefert eine verlässliche kleinräumige Erfassung der Permafrostvorkommen. Elektromagnetische Induktion (wie z.B. mit dem EM-31) kann, sofern sie in Kombination mit zwei-dimensionaler Widerstandstomographie zur Kalibrierung, eingesetzt wird, Permafrostvorkommen in geringer Tiefe über größere Bereiche erfassen. Auf schneebedecktem und überwiegend flachem Gelände kann für größere Areale alternativ der OhmMapper eingesetzt werden.

Introduction

Geophysical methods have been standardly applied in different mountain regions to confirm and characterise mountain permafrost for many years, the basic principle for the successful application being the change of several physical parameters during phase transition from unfrozen to frozen state. In the early years of geophysical investigations of mountain permafrost mainly refraction seismics and direct current (DC) resistivity soundings were used to detect ground ice (e.g. BARSCH 1973, HAEBERLI 1975, FISCH et al. 1977). Later gravimetry and ground penetrating radar complemented the investigation methods (VONDER MÜHLL & KLINGELE 1994, VONDER MÜHLL et al. 2000). During recent years advances have been achieved in using the traditional methods but with more powerful, state-of-the-art instruments and modern data processing algorithms (two-dimensional surveys and data processing, cf. VONDER MÜHLL et al. 2001, HAUCK & VONDER MÜHLL, this volume). Additionally new geophysical techniques were introduced and tested in mountain permafrost studies including electromagnetic induction methods (HAUCK et al. 2001) and recently a further geoelectric method, the so called OhmMapper system (this paper).

The objective of the present work is to test the applicability of different geophysical methods to detect and characterise a sporadic permafrost occurrence below the timberline in the *Bever* valley, eastern Swiss Alps. During the last years several field surveys using different geophysical techniques have been performed at the *Bever* site of which some were applied repeatedly at the same location. At the beginning of the investigation the BTS method (bottom temperature of the winter snow cover, HAEBERLI 1973) has been applied as the standard method for efficiently delineating the distribution of mountain permafrost using BTS probes which were pushed through the snow cover.

The results of the geophysical surveys presented here confirm the existence of low altitude permafrost occurrences, and they provide an opportunity to compare the different methods concerning their applicability on mountain permafrost, which can be a challenge in heterogeneous and steep high mountain terrain. Detailed descriptions of the geophysical methods applied in this study are given for instance in TELFORD et al. (1990) and REYNOLDS (1997).

Study area

The study area is located in the *Bever* valley, a tributary of the main wide valley of the Upper Engadine, eastern Swiss Alps. The *Bever* valley is a trough-shaped valley with bottom elevation between 1730 m and 1800 m a.s.l. at its lower end. The regional climate is rather continental with a fairly low amount of precipitation and a comparatively high temperature range.

At the request of the community of *Bever*, the permafrost distribution in the *Bever* valley and its possible impact on recent and future slope stability and debris flow activity were evaluated (KELLER & KNEISEL 1997). A combination of GIS-based modelling of the potential permafrost occurrence and field investigations were performed. The results of the modelling suggest a potential permafrost occurrence below the timberline on the north exposed valley side, whereas for the south-exposed valley side no permafrost is predicted. On the north facing, orographic right valley side, the lower boundary of the permafrost distribution lies at the foot of the slopes at altitudes between 1800 m and 2000 m a.s.l. On the south exposed, orographic left valley side, permafrost is improbable up to the mountain crest with altitudes between 2700 m to 2950 m a.s.l. (KELLER & KNEISEL 1997). The latter coincides largely with the "rules of thumb" (HAEBERLI 1975) which predict the lowermost permafrost occurrence on south-exposed slopes to be at altitudes above 3000 m a.s.l. It was the first time that a potential permafrost site has been 'discovered' by computer modelling which was confirmed and characterised afterwards by field investigations (KNEISEL et al. 2000).

Both the north- and south-exposed valley sides are wooded. At present the upper timberline is between 2200 m and 2300 m a.s.l. Larch (*Larix decidua*) and cembra pine (*Pinus cembra*) are the dominant tree species of the forests. The steep and rocky valley walls are incised by distinct rock couloirs and torrents, which formed part of the starting zones of several minor debris flows triggered during a thunderstorm in July 1995. Most parts of the scree slopes which occur below the rock walls are well covered with vegetation. The soils are poorly developed and covered by an organic layer of up to 30 cm thickness. Below the organic layer, only a few centimetres of mineral soil exists. The results of the field measurements presented in this paper were concentrated along a clearing on the north-exposed valley side (Fig. 1) where only small larch trees are present.

Data acquisition and data processing

DC vertical electrical sounding

Geoelectrical soundings were applied successfully in different mountain regions to detect mountain permafrost, especially on rock glaciers and (push) moraines (e.g. FISCH et al. 1977, KING et al. 1987, BARSCH & KING 1989, VONDER MÜHLL 1993, EVIN et al. 1997). Due to the noticeable resistivity contrast between unfrozen sediments and ground ice or ice-rich frozen sediments this method is suitable for detecting permafrost.

A typical permafrost sounding curve shows a three-layer model (cf. Fig. 2) with an increase of apparent resistivity at shallow depth representing the active layer) and the permafrost underneath, followed by a sharp decrease of resistivities at greater electrode distances, i.e. greater exploration depths, representing the unfrozen part below the permafrost basis (KNEISEL 1998).

Resistivity values of frozen ground can vary over a wide range depending on the ice content, the temperature and the content of impurities. The dependence of resistivity on

Fig. 1. View of study area in the *Bever* valley with longitudinal geophysical survey line.

temperature is closely related to the unfrozen water content. Perennially frozen silt, sand, gravel or frozen debris with varying ice content show a wide range of resistivity values between 5 kΩm to several hundred kΩm (e.g. HOEKSTRA & MCNEILL 1973, KING et al. 1987, HAEBERLI & VONDER MÜHLL 1996).

Due to the geoelectrical principle of equivalence (e.g. MUNDRY et al. 1985) several calculated models can represent one sounding graph. Thus, it is recommendable to give ranges of resistivity and thickness (VONDER MÜHLL 1993, KNEISEL 1999). For the one-dimensional modelling the software RESIX-PLUS (Interpex) was used. The data presented were obtained using symmetrical Schlumberger and asymmetrical Hummel configurations. Further details of the measurement setup are given in KNEISEL (2002).

DC electrical resistivity tomography

Tomographic resistivity inversions require a large amount of measurements compared to one-dimesional vertical soundings. In order to collect such data sets efficiently, multi-electrode array systems must be employed. We used an ABEM Terrameter SAS 300 C together with an ABEM Lund multi-electrode system for the surveys in 1998 and a SYSCAL Junior Switch system in 2002. Both instruments are quite similar in operation and yield comparable results. For the surveys in 1998 and 2002 we employed 41 (1998) and 36 (2002) electrodes equally spaced at 5 m, which resulted in survey lines of 200 m (1998) and 175 m (2002) length, respectively.

Choice of an appropriate electrode configuration is often dependent on the difficult surface conditions associated with mountain permafrost. Since the maximum current injected into the ground can be quite low as a result of rough surface conditions with high ground resistance and weak signal strength, the geometrical factors of the electrode configurations may be critical (e.g. TELFORD et al. 1990). For this reason, the Wenner configuration may be the best choice for very resistive and heterogeneous terrain because it is less sensitive to weak signal strength, even though Dipole-Dipole configuration may provide superior lateral resolution through a larger number of measurements (LOKE & BARKER 1995, see also HAUCK & VONDER MÜHLL, this volume, KNEISEL, this volume). In order to evaluate the sensitivity of the survey results to the chosen electrode configuration, both Wenner and Dipole-Dipole arrays were tested and compared during the 2002 measurements.

The measured sets of apparent resistivities were inverted using the software package RES2DINV (LOKE & BARKER 1995). This program solves the tomographic inversion problem with a smoothness-constrained least-squares method. A homogeneous earth model is used as the starting model, which is obtained by calculating the logarithmic average of the measured apparent resistivity values.

Capacitive coupled resistivity mapping (OhmMapper)

The OhmMapper (Geometrics) is a capacitively coupled resistivity meter that measures the electrical properties of the subsurface without the galvanic coupling of ground electrodes used in traditional resistivity surveys described above. A simple coaxial-cable array with transmitter and receiver sections is pulled along the ground. Current injection into the ground and the potential measurements are conducted through line antennas placed horizontally on the surface. An alternating current (AC) is induced in the earth at a particular frequency by the alternating voltage applied to the transmitting dipole. The resulting AC voltage coupled to the receiver's dipole is measured. Data is collected in Dipole-Dipole electrode configurations. As with a DC resistivity system, an apparent resistivity is calculated by multiplying the appropriate geometric factor (Dipole-Dipole array) by the OhmMapper's received voltage, normalized by the transmitter current. The operating frequency of the OhmMapper is approximately 16.5 kHz. To obtain a pseudosection as used as input for tomographic inversion with RES2DINV, the survey line must be traversed several times using different spacings. In flat or easy accessible terrain data collection is faster than systems using conventional DC resistivity systems. Further details of this method are given in TIMOFEEV et al. (1994).

Electromagnetic conductivity mapping (EM-31)

Similar to the OhmMapper, the electromagnetic (EM) conductivity meter EM-31 (Geonics) induction technique measures the electrical resistivity, or its reciprocal, the electrical conductivity, without using galvanic coupling with the ground. In contrast to the OhmMapper the EM-31 employs an electromagnetic field at a fixed frequency of 19.8 kHz and a fixed transmitter-receiver spacing of 3.7 m to induce the electrical current in the subsurface. Therefore, there is no need for direct electrical contact between the transmitter and the ground.

It is only recently that these electromagnetic methods have been applied in mountainous regions and especially in the European Alps (SCHMÖLLER & FRÜHWIRTH 1996, HAUCK & VONDER MÜHLL 1999, HAUCK et al. 2001). This is partly due to the fact that apparent resistivities of mountain permafrost bodies are generally much higher than in the Arctic (up to 1 MΩm, compared to 1–10 kΩm in the Arctic) and measured apparent conductivities (1/resistivity) are therefore at the resolution limit of most EM instruments. Nevertheless, conductivity meter like the EM-31 (Geonics) can be applied for conductivity mapping of the uppermost 6 m of the subsurface at great survey speed.

Data acquisition is comparatively easy: measurements can be conducted at walking speed and data processing is done using a simple low-pass filter on the raw data. However, due to the fixed frequency and transmitter-receiver spacing no depth resolution can be obtained (in contrast to EM instruments with variable transmitter-receiver spacing such as the Geonics EM-34).

Refraction seismic tomography

Refraction seismics have a long tradition in permafrost studies (e.g. TIMUR 1968, ZIMMERMAN & KING 1986, KING et al. 1988). Similar to the resistivity investigated by electric methods, the sharp increase of the seismic primary wave (P-wave) velocity at the freezing point is used to differentiate between frozen and unfrozen material. The P-wave velocity distribution can be used as a complementary indicator to resistivity for the presence of frozen material. The method is especially useful to determine the top of the permafrost layer, as the contrast for the P-wave velocity between the unfrozen top layer (=active layer, 400–1500 m/s) and the permafrost body (2000–4000 m/s) is usually large.

A Geometrics EG&S seismograph system with 12 channels was used for seismic data acquisition. The time needed for the completion of one survey line depends nearly linearly on the number of shot points. To allow for acceptable resolution in accordance with synthetic modelling results (cf. HAUCK 2001), two sets of 12 geophones were placed with 5 m spacing. An overlap of 3 geophones and a total number of 10 shot points enabled the data set to be inverted with the tomographic inversion algorithm introduced by LANZ et al. (1998). A sledgehammer was used as source.

As a consequence of low ambient noise levels high signal-to-noise ratios were achieved. The first arrival times of the P-waves were picked manually from the seismograms. Due to the comparatively small number of traveltime data, each trace could be analysed in great detail to minimise picking errors (cf. HAUCK 2001).

Results and discussion

One-dimensional DC resistivity soundings

On the north exposed slope (cf. Fig. 1), a longitudinal as well as a cross profile (with the center point in the middle of the longitudinal profile) were carried out at an altitude of 1830 m a.s.l. in 1996. Three-layer models were calculated which are assumed to represent the most prob-

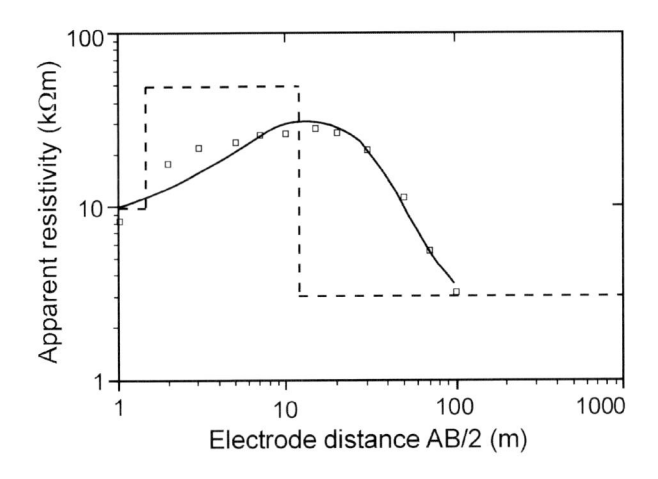

Layer 1		Layer 2		Layer 3
ρ (kΩm)	thickness (m)	ρ (kΩm)	thickness (m)	ρ (kΩm)
8-15	0.5-2	30-90	6-18	2-5

Fig. 2. Results of vertical electrical sounding with ranges of resistivity and thickness for the three-layer model. Dashed line outlines the best matching model.

able case with respect to the local geomorphological situation. The result of the longitudinal profile is displayed in Fig. 2. The shape of the sounding graph is typical of permafrost: increasing resistivities at shallow depth and a distinct decline of the resistivities with larger current electrode distances. A three-layer model inversion was performed with an upper thin layer (approx. 0.5–2 m thick, 10 kΩm) representing the active layer and a middle layer with a thickness of about 10 m representing the permafrost. The apparent resistivities obtained for this middle layer are about 30 kΩm, leading to resistivities between 30–90 kΩm (cf. results of equivalent models shown in Fig. 2). The third layer with resistivities around 3 kΩm can be interpreted as the unfrozen ground underneath the permafrost.

Surprisingly, the active layer seems to be fairly thin. The organic horizons are considered to play an important role in insulating the subsurface controlling the ground thermal regime. This would result in a comparatively thin active layer, as observed in this study. The measurements were repeated in 1998 confirming the results from 1996 (KNEISEL et al. 2000).

Two-dimensional DC resistivity tomography

On heterogeneous ground conditions, the interpretation of one-dimensional soundings can be difficult, as lateral variations along the survey line can influence the results significantly. The sounding curve produces an average resistivity model of the survey area, but individual anomalies as small permafrost lenses will not show explicitly in the results. Two-dimensional resis-

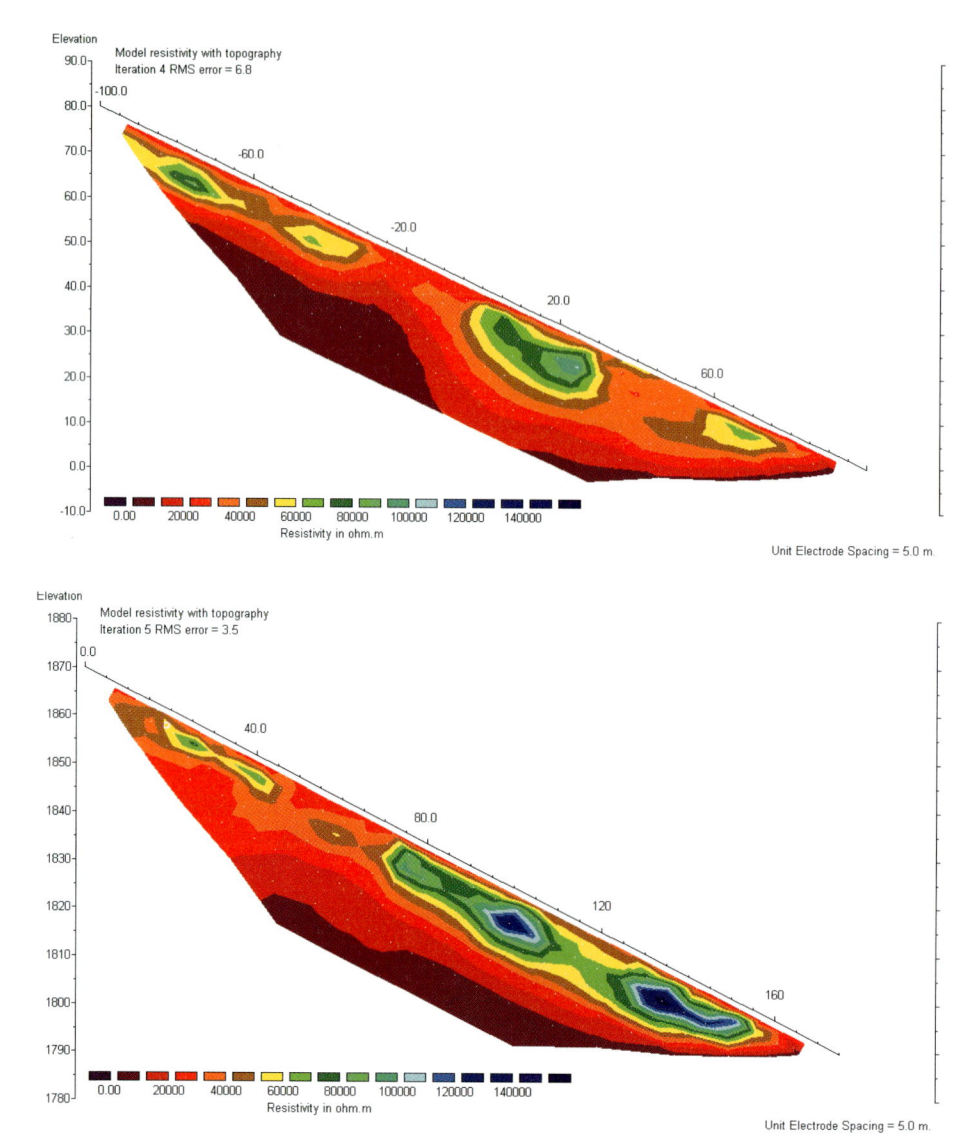

Figs. 3. a and b. Results of the 2D resistivity tomography surveys (Wenner array) in (a) July 1998 and (b) May 2002.

tivity tomography overcomes this problem using multi-electrode systems and two-dimensional data inversion. A combination of both one-dimensional DC resistivity soundings and two-dimensional DC resistivity tomography has proven to allow a more complete characterisation of the sounding site.

The 2D-resistivity surveys (in 1998 and 2002) were conducted at the same location as the 1D-soundings. The model results are displayed in Figs. 3a and b. Resistivity values are as high as 120 kΩm in the uppermost 10 m. Values higher than 20 kΩm, a figure which is considered to indicate permafrost in non-bedrock areas (e.g. HAEBERLI & VONDER MÜHLL 1996), are to be found as deep as 20 m. The lenticular areas of higher resistivity values are interpreted as permafrost lenses and they coincide very well with the findings from the vertical electrical soundings and the refraction seismics (see below).

A comparison between the resistivity surveys conducted in July 1998 and in May 2002 shows very good agreement. The high resistive anomalies are found at the same locations in both cases, with slightly larger modelled resistivity values in the survey of May 2002. This could be due to the slightly different electrode positions, but is more likely a consequence of the fact that the measurements of the year 2002 were performed as early as May. At that time a lesser amount of unfrozen water in the permafrost bodies can be assumed. Furthermore an interannual variation in size and resistivity of the permafrost bodies cannot be excluded at this low altitude permafrost site with aggrading and degrading permafrost depending on the different snow cover history and thermal regime year by year. The thickness of the active layer was less than the resolution limit of the survey geometry resulting from a minimum electrode spacing of 5 m. This also supports the findings from the vertical electrical soundings, which indicated a very shallow active layer of only about 1.5 m thickness for the best model interpretation.

The ice content of the permafrost is difficult to derive from one-dimensional resistivity soundings and two-dimensional resistivity tomography alone. The resistivities are fairly low, which would usually indicate a low ice content. It can be assumed that the permafrost at this low altitude is 'warm', i.e. close to 0 °C, resulting in a high unfrozen water content. This could lead to low resistivities even in the presence of permafrost lenses with rather high ice contents, which could explain the relatively low resistivities found in this study (KNEISEL et al. 2000).

Comparison of electric and electromagnetic methods

Even though electrical resistivity is a very suitable parameter for the detection of permafrost as shown above, there are significant differences between the survey results using various instruments and measurement geometries.

Fig. 4 shows a comparison of the survey results of the various electric and electromagnetic methods used in this study. All results are shown as horizontal variation of electrical resistivity along the longitudinal survey line shown in Fig. 1. In principle, the results of the OhmMapper (a) and the DC resistivity survey with Dipole-Dipole configuration (b) should be the same, as measurements were taken on the same day, at the same location and with the same survey geometry. However, even though the general resistivity distribution is similar for both surveys, the OhmMapper results have a much larger noise content than the DC resistivity data. This is mostly due to difficulties in keeping the correct distance between receiver and transmitter, as well as the correct orientation of the dipole antennas caused by the steep and uneven terrain. No filtering was applied at this stage to be able to compare the raw data sets of both methods. The absolute values of the apparent resistivities are systematically lower for the OhmMapper (1 kΩm) than for the DC resistivity (4 kΩm), which could be due to the high contact resistances between electrodes and ground in DC resistivity surveys on alpine terrain. The question

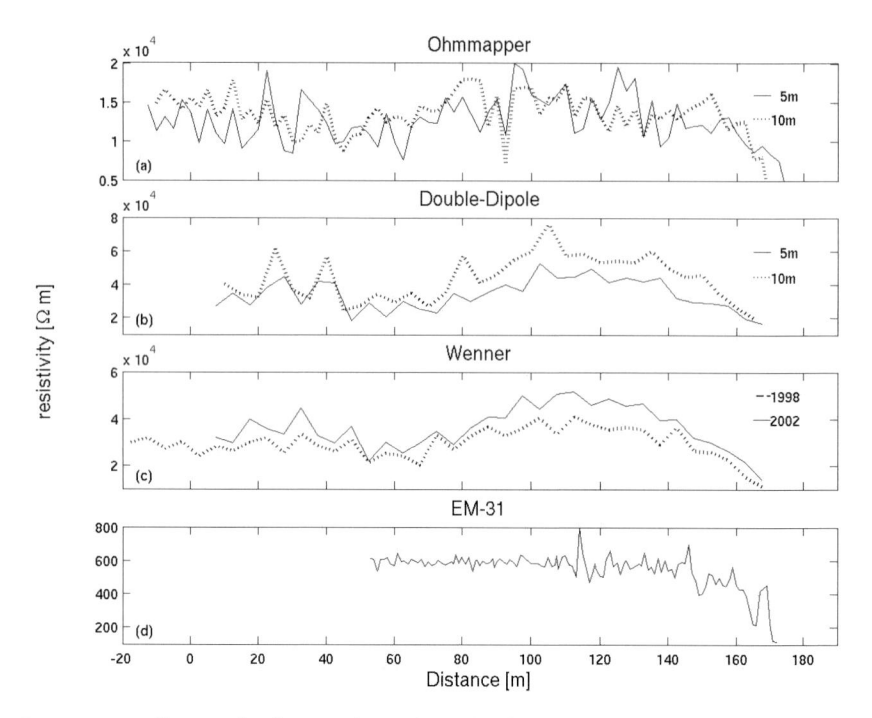

Fig. 4. Comparison of longitudinal survey lines along the clearing conducted with different electric and electromagnetic methods. (a) OhmMapper results from 2002, (b) DC resistivity results from 2002 (5 m and 10 m spacing, Dipole-Dipole), (c) DC resistivity results from 1998 and 2002, (5 m spacing, Wenner) and (d) EM-31 results from 1998.

whether there exists a systematic overestimation of resistivity values resulting from DC resistivity surveys for high-resistive ground conditions is beyond the scope of the present contribution and will be discussed in detail in a separate publication.

Comparing the Dipole-Dipole (Fig. 4b) with the Wenner results (Fig. 4c) from the survey in 2002 with 5 m electrode spacing only small differences are noted. Both, distribution and apparent resistivity values are nearly identical. Furthermore, the systematic positive resistivity difference between the Wenner measurements in May 2002 and July 1998 mentioned above becomes apparent (shown for 5 m electrode spacing).

Finally, the results from the EM-31 survey conducted in July 1998 are shown in Fig. 4d. In comparison to the examples shown in Figs. 4a–c, the resistivity variation along the survey line is markedly different showing almost constant values between horizontal distances 50 and 110 m, in contrast to the resistivity increase in Figs. 4a–c. In addition the apparent resistivity values are more than one order of magnitude less than those of the OhmMapper results, which can be caused by anisotropy, but is more probably partly due to the lack of suitable calibration measurements and partly due to limitations of the sensitivity of the instrument in this high-resistive terrain (HAUCK et al. 2001). However, as a consequence of the large contrast in resistivity between frozen and unfrozen material, the EM-31 can be used for permafrost mapping in combination with a DC resistivity survey. The absolute values obtained with the EM-31 can be "calibrated" using the higher vertical resolution of the DC resistivity results along a common survey line. The much faster survey speed can then be used to map large areas in short time, which would be too time consuming with commonly used DC resistivity techniques (HAUCK & VONDER MÜHLL 1999).

In spite of the inconsistencies mentioned above, electric and electromagnetic methods can be seen as especially well suited to detect permafrost, because the electrical resistivity depends strongly on the unfrozen water content, which is highly dependent on the amount of freezing. However, for certain permafrost environments, interpretation of DC resistivity and EM results are difficult, as the measured resistivity values can be caused by several materials. Whereas the resistivity contrast between ice and unfrozen water is huge, it is small between ice, air and certain rock types, as all three nearly behave as an electrical insulator with very high resistivities. Furthermore, the resistivity values for frozen ground span a wide range from about 10 kΩm to a few MΩm, depending on the ice content.

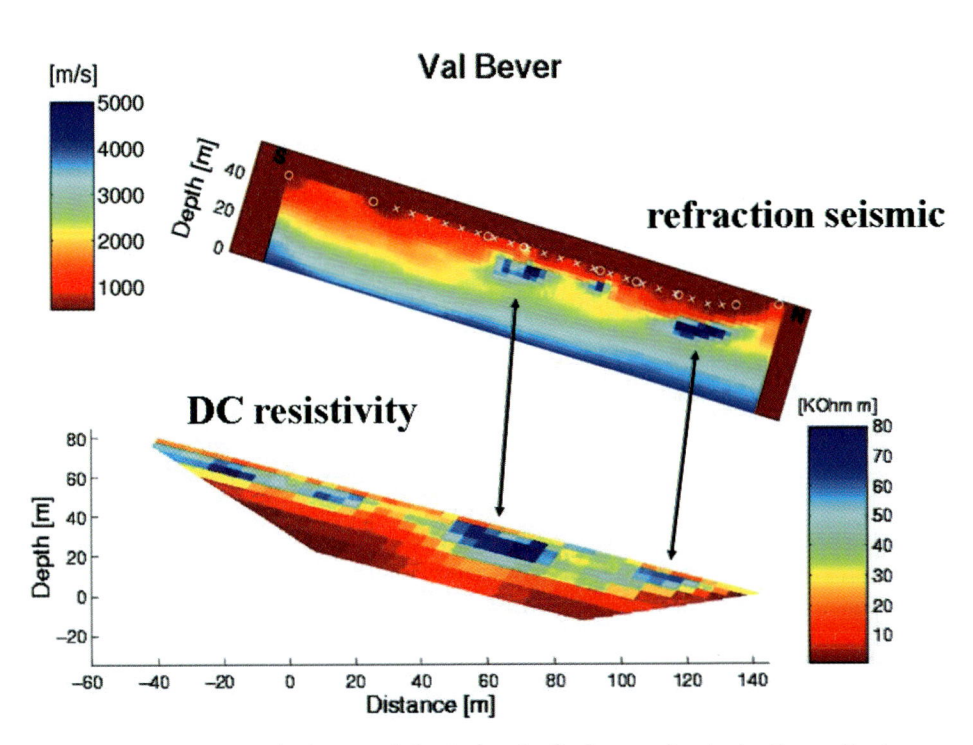

Fig. 5. Seismic inversion results (*top panel*) for the longitudinal survey line in the *Bever* valley in comparison to the DC resistivity tomography results of 1998 shown in Fig. 3 (*bottom panel*). The high-velocity anomalies (2500–4500 m/s, marked with the arrows) in combination with high resistivity values indicate ground ice occurrences in an otherwise unfrozen host material.

Refraction seismic tomography

Seismic P-wave velocities in unconsolidated permafrost are bounded from above (2000–4000 m/s) and the P-wave velocities for ice and air are markedly different (3500 m/s for ice and 330 m/s for air). On the other hand, the values for most common rock types (e.g. sandstone, limestone or granite) are similar to the values for ice, so the detection of mountain permafrost is often difficult by using refraction seismic data alone. From this it becomes clear that for some permafrost applications refraction seismic has to be combined for instance with electric methods, in order to get unambiguous results in terms of permafrost delineation.

The results of the refraction seismic survey line are shown together with the DC resistivity tomography results (1998) in Fig. 5. Isolated anomalies with P-wave velocities between 2500–4500 m/s are present in a host material with velocities between 1000–1500 m/s. The locations coincide with the locations of the high-resistivity anomalies, except that the P-wave velocity anomalies are smaller and much more sharply delineated than the resistivity anomalies. As the Wenner electrode array has comparatively low resolution for detecting laterally varying structures and sharp edged anomalies are smoothed out during inversion (LOKE & BARKER 1995), both survey results represent the same anomalies, which are delineated more clearly by the refraction tomography. Due to the high resistivity values and P-wave velocities between 2500–4500 m/s, they can indeed be characterised as ice lenses, because air cavities would result in much smaller P-wave velocities (cf. HAUCK & VONDER MÜHLL, this volume).

Conclusions

The results of different geophysical techniques applied at the *Bever* site confirmed the existence of permafrost below the timberline. Furthermore, the multi-method approach provides an opportunity for an evaluation of the different methods concerning their applicability on mountain permafrost. Main results from this study include:
- for the described site, the ground thermal regime and thus the permafrost occurrence is assumed to be a result of the interaction of climatic conditions with topography (northern exposure, small amount of incoming radiation, frequent temperature inversions in winter, distribution and duration of snow cover) as well as surface and subsurface factors (organic layers, coarse blocky material). Investigation of the complex ground thermal regime at the Bever site through a near-surface temperature monitoring is in progress and will enable a more complete interpretation of the permafrost occurrence below the timberline.
- the permafrost consists of several thin permafrost lenses at shallow depth (down to 20 m) with a comparatively thin active layer (0.5–1.5 m). Hereby, the organic horizons are considered to play an important role in insulating the subsurface controlling the ground thermal regime. The thickness of the active layer could only be derived by one-dimensional DC resistivity soundings since it was less than the resolution limit of the two-dimensional survey geometry resulting from a minimum electrode spacing of 5 m.
- the detected high resistive anomalies (through one-dimensional DC resistivity soundings and two-dimensional DC resistivity tomography) can be interpreted as frozen ground

since air-filled voids in the vegetated scree-slope, which could also cause high-resistive anomalies, can be excluded through the results of the refraction seismics.
- Electromagnetic conductivity mapping (EM-31) has proven to be a fast method and well suited to locate the horizontal distribution of mountain permafrost at shallow depth, but only in combination with at least one other geophysical method.
- Capacitive coupled resistivity mapping (OhmMapper) allows a faster data collection than systems using a conventional DC resistivity setup, at least in flat or easy accessible terrain, but is also not suited as single method for a comprehensive characterisation of a permafrost site due to the very high noise content.

In general a single geophysical method can lead to ambiguous results concerning the detection and characterisation of permafrost if no additional data from e.g. temperature measurements are available; hence the combination of at least two methods is recommended.

For some geomorphological problems the application of a single geophysical method, such as geoelectrics, might be sufficient. For instance, large air-filled voids, which could be misinterpreted as ice, can be excluded on obvious geomorphological structures such as ice-cored moraines. If any, we currently consider two-dimensional DC resistivity tomography as the most promising single method for detecting mountain permafrost and obtaining a comprehensive characterisation of the subsurface.

Acknowledgements

The study was funded partly through the *Forschungsfonds* at the University of Trier (C. Kneisel) and partly by the PACE project (Contract Nr ENV4-CT97-0492 and BBW Nr 97.0054-1, C. Hauck). C. Hauck acknowledges a grant by the German Science Foundation (DFG) within the *Graduiertenkolleg* Natural Disasters at the University of Karlsruhe.

Many thanks to B. Siemon, A. Hoerdt and an anonymous reviewer for valuable comments on an earlier version of the manuscript.

References

BARSCH, D. (1973): Refraktionsseismische Bestimmung der Obergrenze des gefrorenen Schuttkörpers in verschiedenen Blockgletschern Graubündens, Schweizer Alpen. – Z. Gletscherkd. Glazialgeol. **9**: 143–167.
BARSCH, D. & KING, L. (1989): Origin and geoelectrical resistivity of rock glaciers in semiarid subtropical mountains (Andes of Mendoza, Argentinia). – Z. Geomorph. N. F. **33**: 151–163.
FISCH, W. Sen., FISCH, W. Jun. & HAEBERLI, W. (1977): Electrical soundings with long profiles on rock glaciers and moraines in the Alps of Switzerland. – Z. Gletscherkd. Glazialgeol. **13**: 239–260.
EVIN, M., FABRE, D. & JOHNSON, P. G. (1997): Electrical resistivity measurements on the rock glaciers of Grizzly Creek, St. Elias Mountains, Yukon. – Permafrost Periglac. Process. **8**: 179–189.
HAEBERLI, W. (1973): Die Basis-Temperatur der winterlichen Schneedecke als möglicher Indikator für die Verbreitung von Permafrost. – Z. Gletscherkd. Glazialgeol. **9**: 221–227.
HAEBERLI, W. (1975): Untersuchungen zur Verbreitung von Permafrost zwischen Flüelapass und Piz Grialetsch (Graubünden). – Mitt. Versuchsanst. Wasserbau, Hydrol. Glaziol. ETH Zürich **17**, 221 pp.

HAEBERLI, W. & VONDER MÜHLL, D. (1996): On the characteristics and possible origins of ice in rock glacier permafrost. – Z. Geomorph. N. F., Suppl.-Bd. **104**: 43–57.
HAUCK, C. (2001): Geophysical methods for detecting permafrost in high mountains. – Mitt. Versuchsanst. Wasserbau, Hydrol. Glaziol. ETH Zürich **171**, 204 pp.
HAUCK, C. & VONDER MÜHLL, D. (1999): Detecting Alpine permafrost using electro-magnetic methods. – In: HUTTER, K., WANG, Y. & BEER, H. (eds.): Advances in cold regions thermal engineering and sciences. – pp. 475–482, Springer Verlag, Heidelberg.
HAUCK, C., GUGLIELMIN, M., ISAKSEN, K. & VONDER MÜHLL, D. (2001): Applicability of frequency-domain and time-domain electromagnetic methods for mountain permafrost studies. – Permafrost Periglac. Process. **12**: 39–52.
HAUCK C. & VONDER MÜHLL, D. (2003): Evaluation of geophysical techniques for application in mountain permafrost studies. – Z. Geomorph. N. F., Suppl.-Bd. **132**: 161–190.
HOEKSTRA, P. & MCNEILL, D. (1973): Electromagnetic probing of permafrost. – 2nd Internat. Conf. Permafrost, Proc.: 517–526, Washington D.C.
HOEKSTRA, P., SELLMANN, P. V. & DELANEY, A. (1975): Ground and airborne resistivity surveys of permafrost near Fairbanks, Alaska. – Geophysics **40**: 641–656.
KELLER, F. & KNEISEL, C. (1997): Permafrost im Val Bever. Bericht über die Verbreitung von Permafrost im Val Bever und mögliche Zusammenhänge mit der Rüfentätigkeit. – Unpublished report for the community of Bever.
KING, L., FISCH, W., HAEBERLI, W. & WÄCHTER, H. P. (1987): Comparison of resistivity and radio-echo soundings on rock glacier permafrost. – Z. Gletscherkde. Glazialgeol. **23**: 77–97.
KING, M. S., ZIMMERMAN, R. W. & CORWIN, R. F. (1988): Seismic and electrical properties of unconsolidated permafrost. – Geophys. Prospect. **36**: 349–364.
KNEISEL, C. (1998): Occurrence of surface ice and ground ice/permafrost in recently deglaciated glacier forefields, St. Moritz area, Eastern Swiss Alps. – 7th Internat. Conf. Permafrost, Proc.: 575–581, Yellowknife, Canada.
KNEISEL, C. (1999): Permafrost in Gletschervorfeldern – Eine vergleichende Untersuchung in den Ostschweizer Alpen und Nordschweden. – Trierer Geogr. Stud. **22**, 156 pp.
KNEISEL, C. (2002): Anwendung geoelektrischer Methoden in der Geomorphologie – dargestellt anhand verschiedener Fallbeispiele. – Trierer Geogr. Stud. **25**: 7–20.
KNEISEL, C., HAUCK, C. & VONDER MÜHLL, D. (2000): Permafrost below the timberline confirmed and characterized by geoelectrical resistivity measurements, Bever Valley, eastern Swiss Alps. – Permafrost Periglac. Process. **11**: 295–304.
KNEISEL, C. (2003): Electrical resistivity tomography as a tool for geomorphological investigations – some case studies. – Z. Geomorph. N. F., Suppl.-Bd. **132**: 37–49.
LANZ, E., MAURER, H. R. & GREEN, A. G. (1998): Refraction tomography over a buried waste disposal site. – Geophysics **63**: 1414–1433.
LOKE, M. H. & BARKER, R. D. (1995): Least-squares deconvolution of apparent resistivity. – Geophysics **60**: 1682–1690.
MUNDRY, E., GREINWALD, S., KNÖDEL, K., LOSECKE, W., MEISER, P. & REITMAYR, G. (1985): Geoelektrik. – In: BENDER, F. (Hrsg): Angewandte Geowissenschaften. – Bd. **2**: 299–434.
REYNOLDS, J. M. (1997): An Introduction to applied and environmental geophysics. – Chichester.
SCHMÖLLER, R. & FRÜHWIRTH, R. (1996): Komplexgeophysikalische Untersuchungen auf dem Dösener Blockgletscher (Hohe Tauern, Österreich). – In: Beiträge zur Permafrostforschung in Österreich. – Arb. Inst. Geogr. Karl-Franzens-Univ. Graz **33**: 165–190.
TELFORD, W. M., GELDART, L. P. & SHERIFF, R. E. (1990): Applied geophysics. – 2nd ed., Cambridge University Press.
TIMOFEEV, V. M., ROGOZINSKI, A. W., HUNTER, J. A. & DOUMA, M. (1994): A new ground resistivity method for engineering and environmental geophysics. – Proc. Symp. on the application of geophysics to engineering and environmental problems: 701–715, Boston, Massachusetts.
TIMUR, A. (1968): Velocity of compressional waves in porous media at permafrost temperatures. – Geophysics **33**: 584–595.

Vonder Mühll, D. (1993): Geophysikalische Untersuchungen im Permafrost des Oberengadins. – Mitt. Versuchsanst. Wasserbau, Hydrol. Glaziol. ETH Zürich **122**, 222 pp.

Vonder Mühll, D. & Klingelé, E. (1994): Gravimetrical Investigation of ice-rich permafrost within the rock glacier Murtél-Corvatsch (Upper Engadin, Swiss Alps). – Permafrost Periglac. Process. **5**: 13–24.

Vonder Mühll D., Hauck, C. & Lehmann, F. (2000): Verification of geophysical models in Alpine permafrost using borehole information. – Ann. Glaciol. **31**: 300–306.

Vonder Mühll, D., Hauck, C., Gubler, H., McDonald, R. & Russill, N. (2001): New geophysical methods of investigating the nature and distribution of mountain permafrost with special reference to radiometry techniques. – Permafrost Periglac. Process. **12**: 27–38.

Zimmerman, R. W. & King, M. S. (1986): The effect of freezing on seismic velocities in unconsolidated permafrost. – Geophysics **51**: 1285–1290.

Addresses of the authors: Christof Kneisel, Universität Würzburg, Institut für Geographie, Am Hubland, D-97074 Würzburg. Christian Hauck, Universität Karlsruhe, Institut für Meteorologie u. Klimaforschung, Kaiserstraße, D-76131 Karlsruhe

Evaluation of geophysical techniques for application in mountain permafrost studies

Christian Hauck and Daniel Vonder Mühll

with 2 figures and 3 tables

Summary. An evaluation of geophysical techniques for application in mountain permafrost studies is presented. The applied methods include direct current (DC) resistivity tomography, refraction seismic tomography, electromagnetic induction methods, ground penetrating radar, as well as the classical BTS (bottom temperature of the snow cover) method. The different methods are evaluated concerning their applicability for permafrost detection, vertical and horizontal mapping, monitoring, as well as for active layer studies. The advantages and disadvantages of each method for the respective permafrost problem are discussed in detail.

Zusammenfassung. Die vorliegende Arbeit evaluiert geophysikalische Verfahren in der Permafrostforschung und vergleicht ihre Anwendbarkeit auf typische Fragestellungen. Die getesteten Verfahren beinhalten tomographische Verfahren der Gleichstromgeoelektrik und Refraktionsseismik, elektromagnetische Induktionsmethoden im Frequenz- und Zeitbereich, Bodenradar (GPR), sowie die klassische Methode der BTS (Basistemperatur der Schneedecke). Die Methoden wurden auf ihre Anwendbarkeit zur Auffindung, Charakterisierung, Kartierung, sowie zur Beobachtung zeitlicher Veränderungen von Gebirgspermafrost getestet. Die Vor- und Nachteile der jeweiligen Methoden werden diskutiert.

1 Introduction

Geophysical techniques have been used to study permafrost and characterise areas of permanently frozen ground for many years, but have been mostly applied in polar permafrost regions, where seismic, electromagnetic and electric methods were particularly suitable for exploration and engineering purposes (for a review see SCOTT et al. 1990). Due to the flat and generally uniform surface characteristics in the North-American and Siberian Arctic, the permafrost distribution is quite homogeneous and does not change much on horizontal scales less than a few kilometres. In addition, the surface consists mainly of peat and other materials with a high unfrozen water content. These surface characteristics facilitate the application of most geophysical techniques, since sufficient coupling between sensors and the ground is guaranteed. It is especially important for electric methods, which use direct contact to insert electric current into the ground.

In contrast to this, permafrost occurrences in mountain regions are highly variable and depend strongly on altitude, incoming radiation, local climate and geology. The heterogeneous permafrost distribution calls for methods which are able to resolve the shallow subsurface at scales between a few metres and 1 km. Furthermore, at higher altitudes the surface is often free of vegetation and consists of loose debris with poorly developed soils. Consequently, the unfrozen water content is low, rendering contact of geophysical sensors to the ground more difficult. Surface characteristics are highly variable, thus often excluding the application of commonly used plane-layer approximations for the processing of geophysical data.

Nevertheless, in mountain permafrost the combination of 1-dimensional (1D) DC resistivity soundings and refraction seismics has been applied for many years (BARSCH 1973, FISCH et al. 1977, KING et al. 1987, EVIN & FABRE 1990, KING et al. 1992, VONDER MÜHLL 1993, VONDER MÜHLL & SCHMID 1993, WAGNER 1996). These methods have been particularly effective in determining the depth of the permafrost table and the approximate thickness of the permafrost. In special cases (rock glaciers and ice-cored moraines) the origin of the permafrost ice (sedimentary or congelation/metamorphic, HAEBERLI & VONDER MÜHLL 1996) and/or the ice content could be estimated using DC resistivity and gravimetry (VONDER MÜHLL & KLINGELE 1994). In addition, ground penetrating radar (GPR) was used to map the internal structure of the uppermost 20 to 30 m of rock glaciers (LEHMANN & GREEN 2000, ISAKSEN et al. 2000, BERTHLING et al. 2000). However, most surveys were conducted on special morphologic permafrost features, such as rock glaciers, and only a few geophysical studies have been reported from bedrock permafrost sites.

In order to improve the use of geophysical methods in the study of permafrost the feasibility of various geophysical techniques and state-of-the-art processing schemes for applications in high mountain environments was investigated in a systematic manner during the EU-funded PACE project (Permafrost and Climate in Europe, HARRIS et al. 2001). Applications were not restricted to special morphologies, but covered a broad range of permafrost occurrences in different climatic conditions. To overcome the restriction of plane-layer approximations, 2-dimensional tomographic surveys and processing schemes were used for heterogeneous environments. In order to facilitate geophysical surveys on snow covered, dry and/or debris covered surfaces, the application of electromagnetic induction methods was evaluated, which do not need direct contact between sensors and the ground.

The results are summarised in this contribution. The advantages and disadvantages of each method concerning typical permafrost related questions are presented. Additionally, the characteristics of each method (such as penetration depth, power requirements, persons needed for surveying, data processing) as well as references to case studies in the literature are presented in form of a comprehensive table to facilitate a comparison between the respective techniques. This review should aid the permafrost researcher in the field in choosing the best method for a specific problem.

2 Geophysical methods

2.1 Physical properties of permafrost

Geophysical methods are used to gain information about the physical properties and the structure of the subsurface. In permafrost studies the properties of interest are temperature and ice content. Without a borehole, these properties cannot be observed directly. Therefore, the detection of permafrost from the surface depends on those characteristics that differentiate it from the surrounding media. These are mainly related to changes of physical properties of earth material associated with freezing of incorporated water. The degree of change in the physical properties depends on water content, pore size, pore water chemistry, ground temperature and pressure on the material (SCOTT et al. 1990). The two commonly used geophysical parameters for differentiating between frozen and unfrozen material are the *electrical resistivity* and the *seismic compressional wave velocity*, which is related to the elastic properties of a material.

Electrical resistivity

A large variety of electric and electromagnetic techniques are based on changes of subsurface resistivity. A marked increase in resistivity at the freezing point was shown in several field studies (e.g. HOEKSTRA et al. 1975, SEGUIN 1978, ROZENBERG et al. 1985, BENDERITTER & SCHOTT 1999) and laboratory experiments (OLHOEFT 1978, PANDIT & KING 1978, KING et al. 1988, HAUCK 2001), as shown in Fig. 1a. For many soils, the resistivity increases expo-

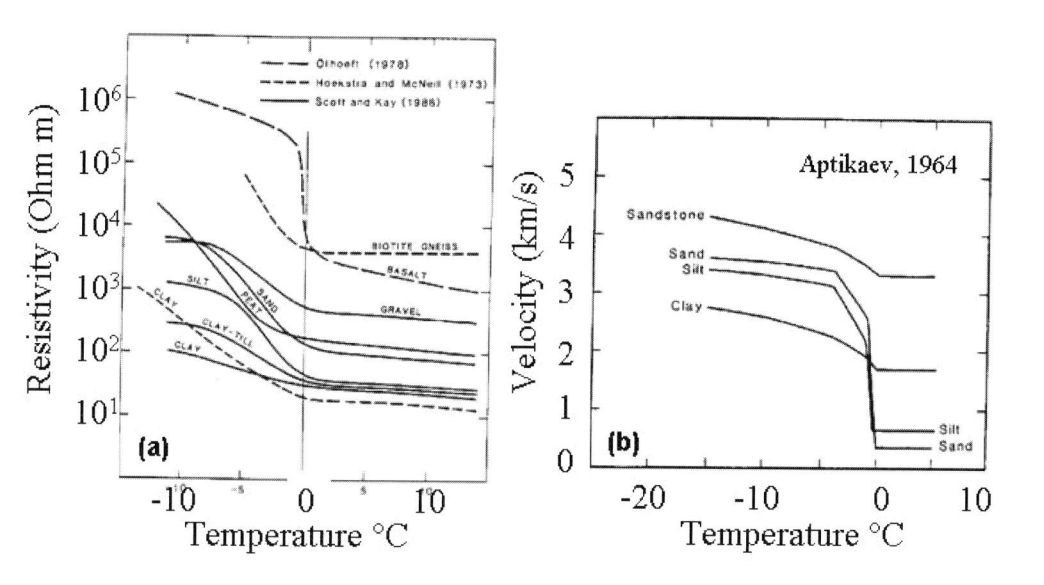

Fig. 1. (a) Resistivity and (b) seismic P-wave velocity of different earth materials as a function of temperature (taken from SCOTT et al. (1990)).

Table 1. Range of resistivities and compressional (or P-wave) velocities for different material types and rocks (compiled after RÖTHLISBERGER 1972, HOEKSTRA et al. 1975, TELFORD et al. 1990).

Material	range of resistivity [Ωm]	range of velocity [m/s]
Air	Infinity	330
Water	$10^1 - 10^2$	1000-1500
Permafrost	polar: $5 \times 10^2 - 10^4$ mountain: $10^4 - 10^6$	2000-4000 (ice: 3500)
Sand	$10^2 - 5 \times 10^3$	400-2000
Clay	$10^0 - 10^2$	500-2400
Sandstone	$5 \times 10^1 - 10^4$	1500-5000
Limestone	$10^2 - 10^4$	3000-6000
Gabbro	$10^3 - 10^6$	3700-7000
Granite	$5 \times 10^3 - 10^6$	3500-5000
Gneiss	$10^2 - 10^3$	4500-6500

nentially until most of the pore water is frozen (McGINNIS et al. 1973, DANIELS et al. 1976, PEARSON et al. 1983, HAUCK 2002). The resistivity is reduced for saline pore waters, as the freezing point is depressed and the unfrozen water content at subzero temperatures is increased (PANDIT & KING 1978). Generally, the resistivity values depend mainly on the material type and the unfrozen water content in the sample (HOEKSTRA & McNEILL 1973, OLHOEFT 1978). A list of resistivity values for common materials is shown in Table 1. Note, that the large range of resistivity values for most materials is due to varying water content (see e.g. Table 5.4. in TELFORD et al. 1990).

Seismic velocity

Seismic techniques make use of changes in the compressional- and shear-wave velocities of rocks and soils. Upon freezing compressional- and shear-wave velocities of most materials increase sharply (see Fig. 1b). This increase is more pronounced the larger the porosity of the material (McGINNIS et al. 1973). An increase in pore-water salinity reduces this effect near the freezing point, as freezing occurs over a finite temperature range instead of at a single temperature (PANDIT & KING 1978). Similar to resistivity, the increase in velocity upon freezing is closely related to the decrease in unfrozen water content (e.g. KING et al. 1988, LECLAIRE et al. 1994). However, in porous media there is an important difference between the behaviour of resistivity and seismic velocities at subzero temperatures. While resistivity continues to increase even at very low temperatures, when the pore space is nearly filled with ice, seismic velocities reach a plateau, where further cooling produces very little change (PANDIT & KING 1978, PEARSON et al. 1983). This illustrates a fundamental difference in the mechanisms by which electrical and acoustic energy are transmitted in rocks. Electric conduction takes place in the unfrozen portion of the pore water, so electrical properties remain sensitive to the amount of unfrozen water present, even if the unfrozen water content becomes very small. In contrast, seismic wave energy is transmitted primarily through the solid matrix, so once the pore volume is largely filled with ice, a further decrease in the already small unfrozen water content produces only a negligible change in velocity (PEARSON et al. 1983). This contrast may also be seen by comparing the range of resistivities and seismic velocities given in Table 1. While the resistivities of permafrost materials are very high with almost no upper bound, the compressional velocity of ice is confined to values around 3500 m/s, which is less than the velocity of most rock types.

2.2 *DC resistivity tomography*

Since a marked increase of the electrical resistivity occurs at the freezing point, electrical methods are expected to be most suitable to detect, localise and characterise permafrost structures. In contrast to most applications of 2-dimensional (2D) and even 3D direct current (DC) resistivity tomography in environmental and engineering studies (e.g. DAILY et al. 1992, GRIFFITHS & BARKER 1993, BINLEY et al. 1996, OGILVY et al. 1999), mountain permafrost targets can be highly resistive (several kΩm to MΩm). In spite of the resulting difficulties in getting sufficient electrical current into the ground the method was successfully applied to map and characterise different mountain permafrost structures in the European mountains (HAUCK & VONDER MÜHLL 1999a, KNEISEL et al. 2000, VONDER MÜHLL et al. 2001a, ISAKSEN et al. 2002, HAUCK 2002, HAUCK et al. 2003, MARESCOT et al. 2003, KNEISEL & HAUCK, this volume) and the Himalaya (ISHIKAWA et al. 2001).

The large number of applications of DC resistivity tomography in recent years was partly due to the availability of 2D inversion software like RES2DINV, which performs smoothness constrained inversion using finite difference forward modelling and quasi-Newton inversion techniques (LOKE & BARKER 1995, 1996). The inversion results in a 2D resistivity model section. In addition, topography may be incorporated in the inversion, which may be an important factor in mountain permafrost terrain.

DC resistivity tomography in 2 dimensions can be conducted using different electrode arrays. The most commonly used geometries are the Wenner, Wenner-Schlumberger and Double-Dipole arrays. A comprehensive evaluation of the characteristics of the specific arrays is given in a resistivity tomography tutorial by LOKE (1999), together with useful information for the conduction of resistivity surveys and data inversion. The penetration depth d is limited by the maximum electrode spacing and may be estimated using a formula by BARKER (1989) with $d = 0.17*L$ for the Wenner array (with L being the distance between the outer (current) electrodes).

2.3 Refraction seismic tomography

Refraction seismic has a long tradition in permafrost studies (e.g. TIMUR 1968, ZIMMERMAN & KING 1986, KING et al. 1988). Similar to electric methods the sharp increase of the P-wave velocity at the freezing point is used to differentiate between frozen and unfrozen material. The P-wave velocity distribution can be used as a complementary indicator to resistivity for the presence of frozen material. The method is especially useful to determine the top of the permafrost layer, as the contrast for the P-wave velocity between the unfrozen top layer (=active layer, 400–1500 m/s) and the permafrost body (2000–4000 m/s) is usually large. The penetration depth depends mainly on the seismic velocity distribution in the ground and the source energy. For most permafrost applications and a sledgehammer as source it is slightly smaller than the corresponding penetration depth of resistivity surveys with similar horizontal survey lengths.

In permafrost studies refraction seismic interpretation techniques are based largely on simple plane-layer models, commonly restricted to two or three layers. This approach may be of limited use for very heterogeneous ground conditions in mountainous environments. As for the DC resistivity technique, tomographic inversion schemes can be used for reliable 2D interpretation (LANZ et al. 1998, HAUCK 2001, MUSIL et al. 2002, SANDMEIER 2002).

2.4 Frequency-domain electromagnetics (FEM)

Similar to DC resistivity techniques, electromagnetic (EM) induction techniques measure the electrical resistivity, or its reciprocal, the electrical conductivity (in Siemens/metre or usually milli-Siemens/metre, mS/m), but with the difference that no direct (galvanic) contact with the ground is needed. This is a great advantage in high mountain environments, as getting sufficient electrical current into the ground is one of the largest problems in DC resistivity surveys. Furthermore, DC resistivity surveys in winter time are usually impossible to conduct, as a dry snow cover acts as an electrical insulator. Electromagnetic induction methods employ a magnetic field to induce the electrical current in the subsurface. Therefore, there is no need for direct electrical contact between the transmitter and the ground (MC NEILL 1980).

In spite of numerous studies concerning permafrost detection conducted in Arctic regions (see SCOTT et al. 1990), it is only recently that these methods have been applied in mountainous regions and especially in the European Alps (SCHMÖLLER & FRÜHWIRTH 1996, HAUCK et al. 1999b, HAUCK et al 2001). This is partly due to the fact that apparent resistivities of mountain permafrost bodies are generally much higher than in the Arctic (up to 1 MΩm, compared to 1–10 kΩm in the Arctic, see Table 1) and measured apparent conductivities (1/resistivity) are therefore at the resolution limit of most EM instruments. Nevertheless, conductivity meter like the Geonics EM-31 can be applied for conductivity mapping of the uppermost 6 m of the subsurface at great survey speed.

Electromagnetic induction is based on the principle that each current-carrying wire is surrounded by circular, concentric magnetic field lines. If bent into a small loop, the wire produces a primary magnetic dipole field. In electromagnetic induction this dipole field is varied, either by alternating the current (operating in the *frequency-domain (FEM method)*) or by terminating it *(transient methods)*, operating in the *time-domain* (TEM, see next section). This time-varying magnetic field induces very small eddy currents in the Earth. The eddy currents generate a secondary magnetic field, which may be sensed at a receiver loop at the surface. The more conductive the subsurface, the larger are the eddy currents and the larger is the measured secondary field, which in turn allows the ground conductivity to be determined by a simple proportional relation. No further data processing is required. Due to the fixed transmitter-receiver distance (3.7 m) the maximum penetration depth is constant (around 6 m, see McNeill 1980).

2.5 Time-domain electromagnetics (TEM)

Contrary to FEM, with TEM instruments the induced secondary magnetic field can be measured by the receiver in the transmitter-off periods, as the primary magnetic field is not alternated but terminated. The response of the subsurface in terms of the decaying amplitude of the secondary magnetic field can then be measured as a function of time and therefore of depth, because later responses originate at greater depths. From repeated measurements a sounding curve similar to an 1D DC resistivity sounding is obtained.

In TEM studies different measurement configurations are utilised. Central loop soundings, where the receiver coil is located in the centre of a large transmitter coil (usually 40 m × 40 m or 100 m × 100 m) have been conducted on permafrost by e.g. Rozenberg et al. (1985) and Todd & Dallimore (1998). Harada et al. (2000) used an outside configuration, where the receiver was placed outside a 60 m x 60 m transmitter loop to reduce the noise level through primary field effects. On mountain slopes or rock glaciers such large transmitter loops are often impossible to use due to the blocky and irregular terrain. Alternatively, a flexible eight-turn 5 m × 5 m transmitter loop can be applied, which allows for reasonably easy handling and a still sufficient penetration depth (Hauck et al. 2001). The maximum depth of penetration can be approximately calculated by

$$h = 8.94 \, l^{0.4} \rho^{0.25},$$

where h is the depth in m, l is the transmitter loop size in m and ρ is the upper layer specific resistivity in Ωm (Geonics 1994). Although the time for one sounding is rather short (a few minutes!) it takes a rather long time for setting up the system. Data processing is conducted analogous to 1-dimensional resistivity inversions (Schlumberger soundings) (e.g. with the Terraplus software TEMIX).

2.6 Ground penetrating radar

Ground penetrating radar (GPR) is a comparatively new geophysical tool, the first instruments being commercially available in the mid 1970's. In permafrost studies they have been successfully used to study the permafrost distribution and structure within the permafrost section (e.g. Annan & Davis 1978, Arcone et al. 1998, Berthling et al. 2000, Vonder Mühll et al. 2000, Lehmann & Green 2000). Similar to the reflection seismic method, an energy pulse is directed into the ground and the arrival times of reflections from subsurface interfaces are recorded. Consequently, data processing is analogous to seismic reflection processing. The central operating frequency can range between 1 to 1000 MHz, and is usually chosen depending on ground attenuation and resolution. Data interpretation is simplest when the ground consists of vertically stacked layers of homogeneous, non-dispersive dielectric materials. Attenuation is highest (and therefore penetration lowest) in low-resistive materials (e.g. materials with a high liquid water content) and particularly in fine-grained sediments (even when frozen), where penetration depths can be less than 1 m (SCOTT et al. 1990). Because of this, and the high dielectric contrast at the frozen surface, GPR is best suited for obtaining detailed information on active layer geometry. Data processing can be rather complex, however as the method works analogous to the well known seismic reflection technique, commercial processing software is readily available (e.g. SANDMEIER 2002).

2.7 Bottom temperature of snow cover (BTS)

The BTS (Bottom Temperature of the Snow cover) method introduced by HAEBERLI (1973), uses the insulation effect of the winter snow cover to the underlying soil. It has been applied as a standard method to delineate the distribution of mountain permafrost in the Alps and Scandinavia for many years (e.g. HOELZLE et al. 1999, ISAKSEN et al. 2002).

When the snow cover is dry and about 1 m or more thick, the temperature at the bottom of the snow cover is effectively shielded from short-term variations in the surface energy balance and remains nearly constant. Then, the BTS is mainly controlled by heat transfer from the upper ground layers, which in turn is strongly dependent on the presence of permafrost. The BTS can be measured using standard thermistor rods. In addition, passive microwave radiometry can be used to measure the BTS on a larger footprint, using the different thermal signatures of snow cover, frozen and unfrozen ground (VONDER MÜHLL et al. 2001a, 2002). A BTS value of less than $-3\,°C$ is generally considered corresponding to a probable permafrost occurrence, a BTS of greater than $-2\,°C$ to a probable non-permafrost site (e.g. HOELZLE 1992, HOELZLE et al. 1999). However, the BTS is not a physical constant but varies from one year to the next. Snow fall history of the winter can modify the BTS markedly (VONDER MÜHLL et al. 1998). Due to the condition of a dry, permanent and at least 1 m thick snow cover, the method is often not applicable on many wind-blown sites, such as exposed mountain tops.

3 Aims in permafrost studies

Permafrost problems amenable to solution by geophysical means fall into three major groups:
- those defining the distribution,
- those defining the temporal evolution and
- those defining the properties.

Permafrost distribution

The problem of defining the permafrost distribution may be subdivided into definition of lateral extent (mapping) and vertical extent (sounding). In addition, determining the extent of the active layer is an important task in many environmental and engineering studies.

Temporal evolution

Determining the temporal evolution of permafrost occurrences by geophysical means has rarely been done in the past, so little information is available yet on the applicability of the different methods. In a changing climate, monitoring of permafrost distribution, and detection of changes in the physical properties of mountain permafrost are among the most important tasks to assess changes in permafrost related natural hazards. Temperature monitoring in deep boreholes and monitoring of BTS temperatures on a larger spatial scale present the foremost tasks to be able to detect permafrost degradation. However, to detect changes in the physical properties, like decreasing ice content or increasing unfrozen water content, monitoring of geophysical parameters is essential.

Permafrost properties

Characterising the permafrost properties, such as ice content and material type, often involves the application of more than one geophysical method, as different permafrost (or non-permafrost) materials may exhibit the same value of a certain geophysical parameter. By using more than one method the uncertainty in the interpretation of the survey results of a single method is greatly reduced.

4 Discussion

In this section each geophysical method is discussed in detail concerning its applicability to the various aspects in mountain permafrost research listed above. Before considering specific permafrost problems, the applicability of the methods for detecting mountain permafrost in general is discussed. Ideally, the presence of permafrost should be detected without using any a priori information, and independent of host material, ice content and dimension of the permafrost occurrence. Advantages and disadvantages of each method will then be evaluated for the following aspects in permafrost studies: spatial mapping, permafrost thickness, active layer studies and permafrost monitoring. The results are summarised in Table 2 and Table 3.

Table 2. Characteristics of geophysical techniques.

Method	Applications (references are given where case studies can be found in the literature)	Persons needed for survey	Comments
DC resistivity tomography	Detection of massive ice in rock glaciers, moraines and other periglacial phenomena (HAUCK & VONDER MÜHLL 1999a, ISHIKAWA et al. 2001, HAUCK et al. 2003, MARESCOT et al. 2003) Mapping isolated ice occurrences (KNEISEL et al. 2000) Monitoring the temporal evolution of permafrost and visualising of transient processes (HAUCK 2002) Laboratory experiments to determine material properties (HAUCK 2001) Determining the altitudinal permafrost limit (ISAKSEN et al. 2002)	1–2	• Obtaining good electrical contact between electrodes and ground is essential • Experience in data inversion is needed for data processing • Differentiation between ice, air and special rock types can sometimes be difficult
EM induction mapping	Mapping isolated ice occurrences (HAUCK & VONDER MÜHLL 1999b) Mapping the boundaries of large periglacial phenomena (e.g. rock glaciers) (HAUCK et al. 2001) Mapping horizontal differences in the active layer thickness (HOEKSTRA & MCNEILL 1973) Determining the amount of heterogeneity to assess the representativeness of single point measurements (HAUCK et al. 2001, HAUCK 2001)	1	• Different surface conditions may have a large influence on the survey results • Instrument drift may lead to erroneous results due to small measurement values • Simple data processing
EM induction sounding	Determining the thickness of a permafrost layer (HARADA et al. 2000, TODD & DALLIMORE 1998, HAUCK et al. 2001)	1–2, 3 for carrying	• No resolution in uppermost 5–10 m • Poor in resolving the exact resistivity value of a resistive middle layer
Refraction seismic tomography	Detection of massive ice in rock glaciers, moraines and other periglacial phenomena (MUSIL et al. 2002) Mapping isolated ice occurrences (KNEISEL & HAUCK 2003, this issue) Differentiation between ice, air and special rock types, each exhibiting anomalously high resistivity values (HAUCK 2001, GUDE et al. 2003) Mapping the active layer thickness	2–3	• Number of receivers should be 12 at least, with shots every other receiver location • Sledgehammer as source is sufficient for most applications • Experience in data inversion is needed for data processing

Method	Applications (references are given where case studies can be found in the literature)	Persons needed for survey	Comments
BTS	Detection and mapping of permafrost occurrences (numerous, e.g. HAEBERLI 1973, HOELZLE et al. 1999, ISAKSEN et al. 2002; concerning the radiometry technique: VONDER MÜHLL et al. 2001a, 2002)	1	• Only applicable if snow cover thickness > 0.8–1.0 m for some weeks • History of snow cover evolution can influence the results • Single point measurements. Representativeness depends on ground surface characteristics
Ground penetrating radar	Delineation of the boundaries of massive ice in rock glaciers, moraines and other periglacial phenomena (LEHMANN & GREEN 2000, ARCONE et al. 1998) Determining the thickness of a permafrost layer (LEHMANN & GREEN 2000) Mapping the active layer thickness (ARCONE et al. 1998)	2–4, depending on the terrain	• Small penetration depth in case of conductive near-surface layers • Not applicable in very heterogeneous media • Experience in data processing is needed

Method	Measured property	Penetration depth	Data processing	Power requirements/logistics
DC resistivity tomography	electrical resistivity	0.17*L (L = current electrode spacing, Wenner array, BARKER 1989)	Software packages available (e.g. RES2DINV) – comparatively easy	Power supply through rechargeable battery packs; the use of spare batteries is recommended
EM induction mapping	electrical resistivity	EM-31: 6 m. other: depending on instrument geometry	none	Small, commercially available batteries for EM-31, rechargeable datalogger
EM induction sounding	electrical resistivity	depending on upper layer resistivity and loop size (see section 2.5)	Software packages available (e.g. Terraplus TEMIX), similar to vertical electrical soundings	Power supply through rechargeable battery packs; the use of spare batteries is recommended
Refraction seismic tomography	seismic P-wave velocity	depending on shot energy and velocity distribution, usually slightly smaller than DC resistivity	First arrival picking. Software packages available (REFLEXW) – some experience needed.	Rechargeable battery for the seismograph
BTS	temperature	none	None	None except for voltmeter
Ground penetrating radar	dielectrical properties, difference in impedancy	depending on attenuation and frequency, usually small for moist subsurface conditions	Software packages available (REFLEXW, SANDMEIER 2002) – experience needed	Power supply through rechargeable battery packs. A laptop computer is needed for surveying.

Table 3. Qualitative comparison of applicability of various geophysical methods.

	PF: yes/no	PF distribution lateral	PF distribution vertical	PF distribution active layer	PF properties	PF monitoring
DC resistivity tomography	applicable	partial	partial	applicable	applicable	applicable
FEM: EM/31,	partial	applicable	non-applicable	applicable	non-applicable	Future studies!
TEM: PROTEM	partial	partial	applicable	non-applicable	partial	Future studies!
refract. seismic tomography	applicable	partial	partial	applicable	applicable	
BTS, radiometry	partial	applicable	non-applicable	non-applicable	non-applicable	applicable
GPR	non-applicable	applicable	applicable	applicable	partial	

■ applicable   □ non-applicable

4.1 DC resistivity tomography

4.1.1 General mountain permafrost detection

Advantages

Judging from the results of a number of studies in recent years the most universally applicable method is the DC resistivity tomography technique. Permafrost was successfully detected for several different permafrost environments including rock glaciers, ice-cored moraines, a low-altitude forested valley slope and bedrock (HAUCK & VONDER MÜHLL 1999a, VONDER MÜHLL et al. 2000, KNEISEL et al. 2000, HAUCK et al. 2000, ISHIKAWA et al. 2001, ISAKSEN et al. 2002, HAUCK et al. 2003, MARESCOT et al. 2003). Because the method primarily depends on the unfrozen water content, it is very sensitive to temperature changes, even in the presence of very high resistivity values. The method is applicable under various surface conditions, provided that there is sufficient electrical contact between electrodes and the ground. In the case of high contact resistances at the electrodes this can be enhanced by using sponges soaked in salt water. Commercially available 2-dimensional inversion software exists (such as RES2DINV) and can be used to process the data in a comparatively simple way.

Disadvantages

Due to the galvanic contact required between electrodes and ground, DC resistivity is usually only applicable on snow-free surfaces. Furthermore, the high contact resistances obtained between electrodes and ground often prohibit the injection of sufficient electrical current into the ground (VONDER MÜHLL 1993). The sensitivity of the inversion results to the input data set is low for highly resistive model regions. Highly resistive anomalies tend to be underestimated by the inversion process (HAUCK 2001, HAUCK et al. 2003). Finally, the applicability of the method depends on the contrast in resistivity between the frozen and unfrozen state of the material. For materials with very low water content, as in many rocks, this contrast can be very small.

4.1.2 Mapping lateral variation of mountain permafrost

Advantages

Due to its 2-dimensional representation, DC resistivity tomography is well suited to determine lateral variations of permafrost. The horizontal resolution can be increased by decreasing the electrode spacing. For data sets with low measurement noise, the representation of small-scale variations can be improved by using smaller regularisation factors in the inversion (HAUCK et al. 2003).

Disadvantages

Even though DC resistivity is a suitable method for delineating permafrost occurrences of limited extent, it is not feasible for mapping large areas. Three-dimensional survey techniques have been proposed recently (e.g. LOKE & BARKER 1996), but survey speed is slow and inversion routines can only process a comparatively small number of data points. Another possibility is to use several 2D survey lines along a 3D grid. However, as a 200 m DC resistivity survey line takes approximately 90–120 minutes to measure, survey areas of one to several square-kilometres are nearly impossible to cover with a reasonable logistic effort. On smaller scales, the smoothing effect of the Wenner array often prohibits the detection of small-scale variability on the order of one to two electrode spacings.

4.1.3 Determining the vertical extent of mountain permafrost (sounding)

Advantages

Determining the vertical extent of permafrost is one of the major tasks in permafrost studies. The classical approach is to use vertical electrical soundings, assuming vertical layering and laterally homogeneous ground conditions. As pointed out above, at mountain permafrost sites this assumption is seldom valid, as ground conditions can be very variable on a small scale. Due to its 2-dimensional representation DC resistivity tomography surveys overcome the problem of lateral heterogeneity. The thickness of the active layer can be determined and for shallow permafrost occurrences the depth of the permafrost base can be delineated. To improve vertical resolution the Wenner-Schlumberger array can be used instead of the Wenner array.

Disadvantages

In order to obtain deep sounding data, large electrode spacings have to be used. To obtain a good vertical resolution, the horizontal resolution has to be increased as well, leading to a large number of electrodes. Consequently, survey speed would be low and data processing requires large computer capacities. Synthetic modelling studies have shown that DC resistivity surveys tend to underestimate the thickness of a resistive middle layer (HARADA et al. 2000).

4.1.4 Active layer studies

Advantages

By reducing the electrode spacing, the active layer can be investigated in great detail. Active layer processes like freezing and thawing can be visualised, through repeated surveys along the same lines (HAUCK 2002). Due to its high sensitivity to unfrozen water content, resistivity monitoring of phase transitions and of the flow paths of rain and melt water is highly effective (HUBBARD et al. 1998, FRENCH et al. 2002).

Disadvantages

In order to resolve the active layer the electrode spacing has to be reduced to at least 2 m. However, decreasing the electrode spacing decreases the penetration depth for a constant number of electrodes (for 41 electrodes with 2 m spacing the penetration depth is around 13 m (after BARKER 1989)). Consequently, simultaneously resolving the active layer *and* the permafrost body below requires a large number of electrodes and increases measurement and processing time.

4.1.5 Permafrost monitoring

Advantages/Disadvantages

In HAUCK (2002), results from repeated DC resistivity tomography measurements using a buried, fixed-electrode array are presented. Resistivity variations in the uppermost 10 m could be related to temperatures and the evolution of the unfrozen water content with time was determined. This approach is probably the most promising for future studies, especially in combination with energy balance measurements and/or other geophysical measurements on a seasonal basis (e.g. seismics). However, care has to be taken in differentiating between resistivity variations due to freeze and thaw processes and variations due to additional water input through precipitation or ground water variability.

4.2 FEM methods

4.2.1 General mountain permafrost detection

Advantages

Conductivity meter are generally light-weight and easy to handle instruments. Data acquisition is fast and data processing is simple (see MCNEILL 1980). As a magnetic field is used to induce the electric current in the ground, no direct contact between the instrument and the ground is needed. Consequently, surveys can be made on snow-covered ground, providing the snow cover is dry and homogeneous.

Disadvantages

Most conductivity meter like the Geonics EM-31 have a fixed transmitter-receiver spacing which prohibits significant depth resolution. Maximum penetration depth is usually limited to about 6 m. On the other hand, FEM instruments with variable transmitter-receiver spacing (such as the Geonics EM-34) are much more difficult to apply on permafrost terrain, as an exact control of the spacing has to be maintained during the survey. Due to the very low conductivities observed in mountain permafrost environments, the conductivity values are close to the resolution limit of most FEM instruments and cannot be used as an indicator for the presence of permafrost (HAUCK & VONDER MÜHLL 1999b). Furthermore, the instrument drift can be substantial compared to the permafrost signal and conductivity anomalies due to changing surface characteristics can mask any conductivity variations in the subsurface (HAUCK et al. 2001). Consequently, reliable detection of mountain permafrost by an EM-31 survey without additional information is limited to some special cases.

4.2.2 Mapping lateral variation of mountain permafrost

Advantages

Due to their mapping speed FEM instruments like the EM-31 are much better suited for covering large areas than 2-dimensional tomographic methods, even though only the average conductivity of the uppermost 6 m is measured and no vertical resolution can be obtained. In HAUCK & VONDER MÜHLL (1999b), a procedure was proposed, combining the advantages of the high resolution DC resistivity tomography method with the mapping speed of the EM-31. By this method, a survey area of several square-kilometres can be investigated in one day.

Disadvantages

For long survey duration and small conductivity contrasts, the instrument drift may influence the survey results (HAUCK et al. 2001). Repeated measurements of survey lines and recalibration at chosen grid points may overcome this problem. On snow-covered ground, influences through variations in snow height and composition have to be taken into account, especially in the presence of wet snow. Due to these constraints, the authors found that FEM mapping of mountain permafrost variability can not be done without the use of additional information at certain representative locations.

4.2.3 Determining the vertical extent of mountain permafrost (sounding)

Advantages/Disadvantages

With the FEM instruments used in this study no significant vertical resolution can be obtained. As described in MCNEILL (1980), some information about vertical conductivity changes may be obtained by using different instrument heights above the ground, different instrument polarisations or different transmitter-receiver spacings. The last option can not be done with the EM-31, but is possible for the Geonics EM-34, which uses two separate coils joined by a flexible cable. However, for deep electromagnetic soundings, TEM methods are much superior to FEM systems (TELFORD et al. 1990).

4.2.4 Active layer studies

Advantages

Generally, FEM methods have a strong applicability to active layer studies, and especially to the lateral variability (HOEKSTRA 1978). Due to its small penetration depth, the EM-31 is very sensitive to changes in the active layer. In contrast to DC resistivity and refraction seismics, FEM methods are easy to use on blocky terrain like rock glaciers, as they do not need electrical contact with the surface. Survey lines can easily be repeated due to the fast survey speed.

Disadvantages

As mentioned above, without using different polarisations or instrument heights, only the bulk conductivity of the active layer can be obtained. Consequently, it cannot be determined whether measured apparent conductivity variations are due to conductivity variations within the active layer and or due to variations in active layer thickness.

4.2.5 Permafrost monitoring

FEM methods are not as feasible for monitoring purposes as the DC resistivity approach described above, because no fixed measurement setup can be installed to accurately monitor conductivity changes. However, as FEM surveys are fast and can be conducted with minimal effort, measurements along the same survey lines may be repeated on a regular basis. Due to the strong sensitivity to instrument height and surface conditions, care has to be taken to ensure similar measurement conditions.

4.3 TEM methods

4.3.1 General mountain permafrost detection

Advantages

Results from PROTEM soundings at various field sites have shown that deep permafrost occurrences can be detected with this method (Rozenberg et al. 1985, Todd & Dallimore 1998, Harada et al. 2000, Hauck et al. 2001). As the TEM methods work in the time-domain, large penetration depths can be obtained with comparatively small measurement geometries at the surface. This is especially useful for deep soundings on geomorphological features with small lateral extent, e.g. rock glaciers and moraines. Similar to the FEM methods, no direct contact between instrument and ground is needed, facilitating the conduction of surveys on snow covered ground.

Disadvantages

Due to technical limitations usually no information about the uppermost 5–10 m can be obtained. Furthermore, spatial resolution is limited to the transmitter coil dimensions (Geonics 1994). Even though TEM methods are highly sensitive to the thickness of a resistive layer, the sensitivity is poor for determining its resistivity value (Maier et al. 1995). Finally, as for all methods utilising electrical resistivity as geophysical parameter, a significant contrast between the resistivity of unfrozen and frozen material is needed to successfully delineate the permafrost occurrence.

4.3.2 Mapping lateral variation of mountain permafrost

Advantages

A series of PROTEM soundings along a profile line or grid may be combined to give a 2D- or even 3D-representation of the permafrost distribution. This approach has been successfully used on polar permafrost sites, where topographical influences and variations in the subsurface geology were small (e.g. Todd & Dallimore 1998, Harada et al. 2000).

Disadvantages

As no information about the uppermost 5–10 m can be obtained and spatial resolution is limited by the transmitter coil dimensions, the application of PROTEM surveys for mapping lateral variability of mountain permafrost is limited. Furthermore, 2D and 3D inversion algorithms are not as readily available as for DC resistivity or seismic techniques. Consequently, most 2D TEM data is shown in the form of pseudosections, which may not represent a reliable 2-dimensional image of the ground, as no data inversion is performed.

4.3.3 *Determining the vertical extent of mountain permafrost (sounding)*

Advantages

For delineating the base of deep permafrost occurrences transient electromagnetic soundings, as performed with the PROTEM was found to be the most promising technique. Measurement speed is fast and the equipment may be carried by 2–3 persons. The size of the measurement configuration is small compared to the penetration depth, yielding penetration depths of more than 100 m with a 5 m by 5 m multi-turn cable. For comparison, a DC resistivity sounding with a Schlumberger array would need a cable spread, and therefore homogeneous ground conditions, of at least 0.5–1 km for a similar penetration depth. The results of the PROTEM soundings were best for targets with large resistivity contrasts, e.g. Alpine rock glaciers (HAUCK et al. 2001).

Disadvantages

The TEM method is poor at resolving the resistivity of an intermediate resistive layer, where DC resistivity methods are much superior. Consequently, joint interpretation or joint inversion of both data types would be very effective at reducing the deficiencies of either technique when used alone (SANDBERG 1993, MAIER et al. 1995).

4.3.4 *Active layer studies*

As explained in section 2.5, the PROTEM transmitter current cannot be stopped instantaneously and the first time gate recorded must occur a certain time after the termination. Consequently, no information from the topmost 5–10 m can be obtained (GEONICS 1994).

4.3.5 *Permafrost monitoring*

In a warming climate, permafrost is likely to thaw from the top down, and the most drastic changes may be seen at shallow depth. As TEM methods give no reliable results for the uppermost 5–10 m, their monitoring potential is believed to be small.

4.4 *Refraction seismic tomography*

4.4.1 *General mountain permafrost detection*

Advantages

Similar to DC resistivity, refraction seismic tomography is well suited to detect mountain permafrost. Logistical effort and survey time are similar for both methods. In contrast to the electric and electromagnetic techniques, refraction seismics make use of a different geophysical parameter, the P-wave velocity (section 2.1). The method provides more structural information about the subsurface than DC resistivity and inductive EM methods (cf. HAUCK 2001, KNEISEL & HAUCK, this volume). Layer boundaries and isolated objects are more sharply delineated in seismic surveys, because the refracted waves travel along boundaries of materials with different elastic properties.

Disadvantages

The potentially higher resolution in seismic surveys is in contrast to the partly lower accuracy of the inversion model results, especially in model regions where ray coverage is poor during inversion (LANZ et al. 1998, HAUCK 2001, MUSIL et al. 2002). The accuracy of the inversion results is given by the number of ray paths crossing each model cell. In model regions with poor ray coverage the resulting P-wave velocity is poorly constrained by the data and the accuracy is low (LANZ et al. 1998). Furthermore, the P-wave velocity of high-velocity anomalies tends to be overestimated by the inversion process (HAUCK 2001). Finally, the P-wave velocity is not as sensitive to temperature changes in partly frozen materials as resistivity, because seismic waves travel along the frozen part of the material, as opposed to the resistivity, which depends on the unfrozen part. In partly frozen material, a further temperature decrease will significantly change the resistivity, whereas the P-wave velocity will not change much, as soon as a continuous frozen matrix has been established.

4.4.2 Mapping lateral variation of mountain permafrost

Advantages

Similar to DC resistivity, refraction seismic tomography is well suited to determine lateral variations of permafrost due to its 2-dimensional representation. The lateral resolution of refraction seismics is high, providing the ray coverage is sufficient. Resolution can be further increased by decreasing the geophone spacing or increasing the number of shot points. For data sets with a low amount of measurement noise, the representation of small-scale variations can be improved by using smaller regularisation factors in the inversion (HAUCK 2001).

Disadvantages

Like DC resistivity, refraction seismic is not feasible for mapping large areas due to logistic requirements. A large number of shot points are needed to obtain a good ray coverage. Data processing is more time consuming than for DC resistivity, because the first arrival times have to be picked prior to the inversion process.

4.4.3 Determining the vertical extent of mountain permafrost (sounding)

Advantages

The thickness of the active layer and the depth of the permafrost base can be delineated for shallow permafrost occurrences. The 2-dimensional representation increases the information content and the reliability of the interpretation. In case of vertical layering and increasing velocities with depth, the depth of the respective layers can be determined with great accuracy.

Disadvantages

In addition to the logistical problems associated with long survey spreads and a large number of sources and receivers, seismic soundings in permafrost terrain are hampered by the difficulty in resolving a large low-velocity zone below a high-velocity zone. If the high-velocity zone, corresponding to the ice-rich permafrost layer, is laterally extensive and velocity decreases with depth, all wave energy will become focused in this layer and no information about deeper, unfrozen layers can be obtained. Furthermore, ray density decreases with depth leading to uncertainties in the thickness of isolated high-velocity anomalies, as shown in model studies with synthetic data sets (HAUCK 2001).

4.4.4 Active layer studies

Advantages/Disadvantages

Due to the high lateral resolution, refraction seismic tomography is very suitable for investigating the active layer in detail. In the shallow subsurface the ray coverage is large and high accuracy can be obtained without reducing the geophone spacing. However, because of the usually large seismic velocity contrasts between active layer and permafrost body, high-resolution surveys with tomographic inversions are usually not needed for getting reliable results. Seismic refraction surveys using single channel seismographs have shown good results concerning the delineation of variable active layer thicknesses (HARRIS & COOK 1986)

4.4.5 Permafrost monitoring

In principle, refraction seismics may be used for monitoring similar to the DC resistivity approach shown in HAUCK (2002). A fixed geophone array can be installed and measurements may be conducted along the survey line using different source points. Determination of velocity changes due to freezing and thawing would facilitate the interpretation of refraction seismic survey data from mountain permafrost areas. Together with changes in resistivity, these velocity changes could be used to accurately determine changes in unfrozen water and ice content. However, in contrast to electrodes, which may consist of cheap steel rods (e.g. tent pegs), it is generally not feasible to leave geophones at fixed locations in the field for a long time. Consequently, difficulties arise in ensuring that the same source and receiver locations are used for repeated surveys.

4.5 Ground penetrating radar (GPR)

4.5.1 General mountain permafrost detection

Advantages

The great survey speed, the high resolution and its ability to determine the interface between unfrozen and frozen material are the great advantages of the GPR technique. This method is especially well suited for delineating internal structures in permafrost bodies, and for application on snow-covered terrain.

Disadvantages

The method is unsuitable if the subsurface is very heterogeneous, because multiple scattering of the electromagnetic waves will render data processing more difficult. The penetration depth is small for fine-grained material and/or wet surface and subsurface conditions. The fibre-optic cables between antennas and control device are very sensitive and must be handled with great care, which is often difficult in mountain terrain. Finally, data processing is comparatively difficult and artefacts may develop through inadequate filtering.

4.5.2 Mapping lateral variation of mountain permafrost

Advantages

Due to its fast survey speed and the large difference between the dielectric properties of unfrozen and frozen material, the GPR technique is very well suited for permafrost mapping of large areas if the depth penetration is sufficient (ARCONE et al. 1998, and references herein).

Disadvantages

The penetration depth is often limited by wet subsurface conditions, as high attenuation rates in water saturated material prohibit the penetration of electromagnetic wave energy to greater depths. This may be overcome to a certain degree by lowering the midband radar frequencies (e.g. 50 MHz, see ARCONE et al. 1998). In addition, difficulties may arise in heterogeneous ground conditions, where scattering losses due to intrapermafrost diffractions occur. This phenomenon can be most pronounced on rock glaciers with irregular internal structures and without a pronounced ice body (MUSIL et al. 2002).

4.5.3 Determining the vertical extent of mountain permafrost (sounding)

Advantages/Disadvantages

As stated above, depth penetration is often limited, which may prohibit significant vertical resolution. However, for large subsurface ice bodies with low unfrozen water contents the penetration depth can be large and the vertical extent can be detected quite accurately using modern processing routines (e.g. migration, LEHMANN & GREEN 2000).

4.5.4 Active layer studies

Advantages/Disadvantages

The high resolution in the uppermost subsurface layer obtained by the GPR method makes it an ideal tool for delineating the geometry of the active layer. Even though many active layer studies have been performed in Arctic permafrost regions (e.g. DOOLITTLE et al. 1990), up to now no active layer studies with GPR have been conducted on mountain terrain to our knowledge.

4.5.5 Permafrost monitoring

As with FEM methods, GPR is not as feasible for monitoring purposes as the DC resistivity approach described above, because no fixed measurement setup can be installed to accurately monitor changes in e.g. active layer thickness. However, as GPR surveys are fast and can be conducted with minimal effort, measurements along the same survey lines may be repeated on a regular basis.

4.6 BTS

4.6.1 General mountain permafrost detection

Advantages

The BTS method directly measures the temperature of the ground and does not have to be related to an additional geophysical parameter. The method is fast, easy to use and no processing of the data is required. Furthermore, BTS measurements have been conducted in mountain permafrost regions since 30 years (especially in the Alps and Scandinavia) with a large amount of literature available (e.g. HOELZLE et al. 1999, and references herein).

Disadvantages

The largest restriction on the applicability of the BTS method is the requirement of an undisturbed snow cover of at least 0.8 – 1.0 m thickness. The snow cover has to be present for about one month prior to the measurements in order that BTS can be related to the permafrost conditions (HOELZLE et al. 1999). In addition, the BTS is not constant from one winter to the next, depending on the snow fall history during the winter (VONDER MÜHLL et al. 1998). Conventional measurements with probes are conducted at individual points underneath the snow cover, depending on the surface characteristics. Furthermore, the method and the commonly used BTS ranges ($>-2\,°C$ and $<-3\,°C$) have been calibrated based on measurements in the Central Alps and a generalisation to other mountain permafrost environments has to be assessed for each region separately (e.g. ISAKSEN et al. 2002).

4.6.2 Mapping lateral variation of mountain permafrost

Advantages

In winter time, the BTS method is still the most important survey technique for mapping the permafrost distribution. It is the only method which determines the permafrost distribution from its thermal properties, without using resistivity or seismic velocity as reference parameter. The survey speed is very fast and no data processing is required. The passive microwave radiometry method introduced in VONDER MÜHLL et al. (2001a) aims to improve this approach in terms of measurement speed, e.g. through airborne measurements, and concerning spatial representativeness. When averaged over larger footprints, the BTS results may lose its dependency on local heterogeneities.

Disadvantages

Because a thermal signal is measured, one cannot distinguish between different subsurface materials or material properties. As explained above, BTS surveys require a significantly thick (0.8–1.0 m) and permanent snow cover for at least one month and the results depend on the snow fall history and may vary from one year to the next. These constraints require careful interpretation of survey results. Surveys should be repeated throughout the winter and in concurring years. Ideally, miniature temperature dataloggers should be used for continuous temperature measurements throughout the year (HOELZLE et al. 1999).

4.6.3 Determining the vertical extent of mountain permafrost (sounding)

No depth resolution can be obtained with this method.

4.6.4 Active layer studies

BTS studies can only be conducted in winter, when the active layer is frozen. However, relationships between BTS, snow height and active layer thickness in summer may be formulated (KELLER & GUBLER 1993).

4.6.5 Permafrost monitoring

Except for DC resistivity monitoring test studies, BTS and borehole temperature monitoring are presently the only methods in operational use for continuous permafrost monitoring. During the pilot phase of the Swiss Permafrost Monitoring Network (PERMOS) BTS measurements are repeated every year at selected test sites in the Alps in order to monitor changes in the permafrost distribution and its thermal regime (cf. VONDER MÜHLL et al. 2001b).

4.7 Reducing uncertainty by using more than one method

Electric and electromagnetic methods are especially well suited for detection of permafrost, because electrical resistivity depends strongly on the unfrozen water content, which is in turn highly dependent on the amount of freezing. However, for certain permafrost environments interpretation of DC resistivity and EM results are difficult, as the measured resistivity values can be caused by several materials. Whereas the resistivity contrast between ice and unfrozen water is huge, it is small between ice, air and certain rock types, as all three nearly behave as electrical insulators with very high resistivities (Table 1). Furthermore, the resistivity values for frozen ground span a wide range from about 10 kΩm to a few MΩm, depending on the ice content. In contrast, seismic P-wave velocities in unconsolidated permafrost are limited to 2000–4000 m/s and the P-wave velocities for ice and air are markedly different (Table 1). On the other hand, the values for most common rock types (e.g. sandstone, limestone or granite) are similar to the values for ice, so the detection of mountain permafrost is often difficult by refraction seismic alone. From this it is clear that for some permafrost applications both methods have to be combined in order to get unambiguous results in terms of permafrost delineation.

In KNEISEL & HAUCK (this issue), a field case is presented where DC resistivity measurements predicted shallow permafrost occurrences, and were verified by refraction tomography results (cf Fig. 5 in KNEISEL & HAUCK, this issue). In contrast, another case study is presented here, where similarly high resistivities were not caused by a permafrost occurrence. For the permafrost researcher using DC resistivity, the main goal is to decide if high resistivity anomalies correspond to air-filled cavities, ice, or bedrock anomalies.

Juvvasshoe, Central Norway: air cavern anomalies

As an opposing example to the case study cited above the following case study addresses the problem of high resistivity values due to air-filled cracks or cavities, which may be misinterpreted for permafrost occurrences. Cracks or air-filled cavities can be found frequently in mountainous terrain, especially near large boulders and underneath a debris covered surface. Resistivities can be anomalously high, as air acts as a perfect electrical insulator. However, the seismic P-wave velocity in air is very low (330 m/s), therefore easy to contrast to the much higher velocities of rock and ice (e.g. GUDE et al. 2003).

The example shown in Fig. 2 originates from the northern slopes near Juvvasshoe, Central Norway. As explained in detail in HAUCK et al. (2000) and ISAKSEN et al. (2002) an extensive survey along the slope below the PACE drill site at Juvvasshoe was conducted to delineate the transition area between permafrost and non-permafrost. A 640 m long DC resistivity profile showed a clear transition zone between high resistivities of up to several tens of kΩm and resistivities as low as 1 kΩm. However, anomalously high resistivity values were also found near the surface in the presumed non-permafrost areas (see dark shaded region *I* in the DC resistivity tomogram in Fig. 2). To determine the cause of these anomalies, a refraction survey was conducted along the same profile line. From the results in Fig. 2 it is clearly seen that air-filled cavities have to be present, as the P-wave velocities in the uppermost 1–2 m are lower than

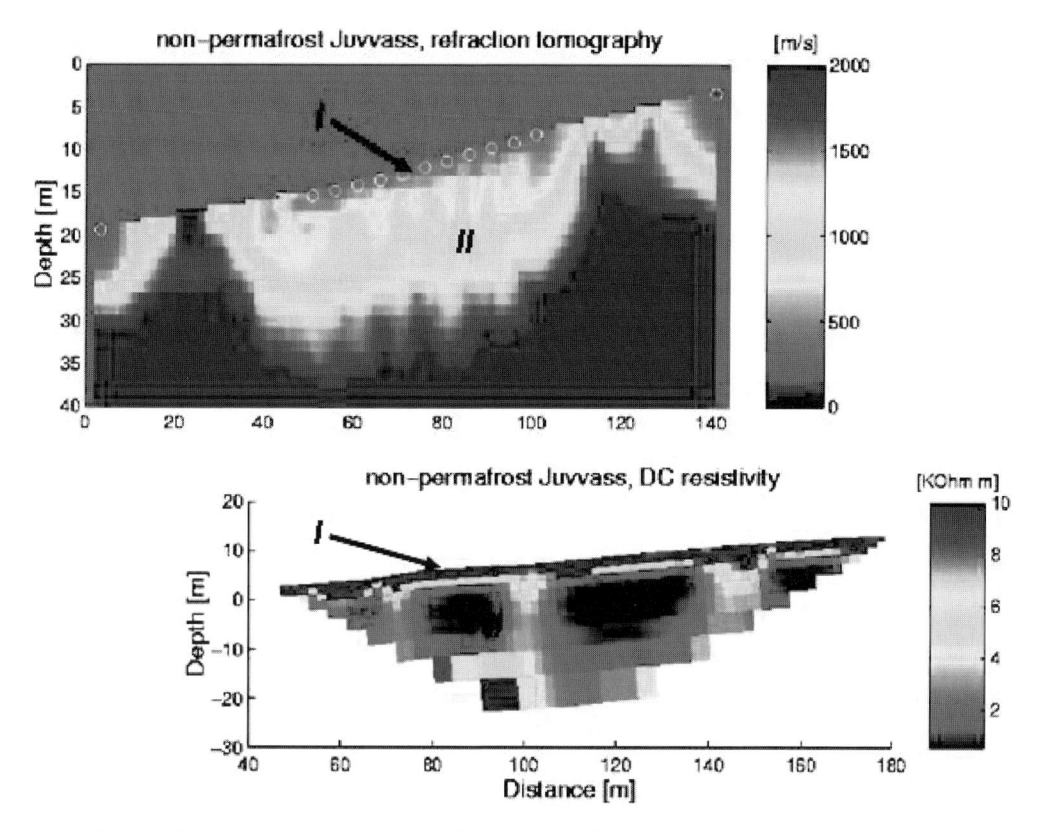

Fig. 2. Seismic inversion results (*top panel*) for the survey line at the non-permafrost site near Juvvasshoe, Norway, in comparison to the DC resistivity tomography results (*bottom panel*). The very low velocities (<500 m/s) in the uppermost 1–2 m indicate the presence of air-filled cavities. Regions *I–II* correspond to features explained in the text.

500 m/s. Below, the values increase slowly to 1000–1500 m/s, indicating the presence of unconsolidated, unfrozen material and water, which is in good agreement with the low resistivity values found at this depth (region *II* in Fig. 2).

5   Conclusions

Applications of a number of different geophysical techniques in mountain permafrost regions were evaluated, using data obtained during numerous field surveys within the PACE (Permafrost and Climate in Europe) project. The aim was to determine suitable methods to detect, characterise and map permafrost at a number of test sites including pure bedrock sites as well as unconsolidated permafrost sites like rock glaciers or moraines. The applied geophysical techniques included DC resistivity tomography, refraction seismic tomography, electromagnetic induction methods, ground penetrating radar (GPR) and the BTS method (bottom temperature of snow cover).

DC resistivity tomography turned out to be a suitable method for a number of permafrost related questions, such as detection of permafrost, mapping of horizontal extent, estimation of ice/unfrozen water content and determination of the permafrost base for shallow permafrost occurrences. Refraction seismic tomography is equally well suited for these targets, except that interpretation of the results is more difficult, as the difference between the measurement signals of frozen and unfrozen ground is smaller than with DC resistivity. Ideally, a combination of both methods should be applied.

Measurements with electromagnetic induction methods included conductivity mapping with conductivity meters (EM-31) and transient electromagnetic soundings (PROTEM).

Due to the large resistivity values encountered and the heterogeneity of most permafrost field sites in high mountain environments, conductivity mapping using the EM-31 has to be combined with other geophysical techniques, like DC resistivity. Once a permafrost occurrence is detected by a DC resistivity survey its extensions can easily be mapped with a conductivity meter. Electromagnetic soundings using the time-domain instrument PROTEM can be used to determine the permafrost base. The maximum penetration depth depends on the transmitter coil dimensions and the upper layer resistivity. Ground penetrating radar is best for determining the internal structure of permafrost bodies, as well as for detecting the spatial extent of individual layers. However, in case of a conductive overburden, penetration depth can be shallow.

If possible several geophysical techniques should be used together, as results from only one method may yield ambiguities concerning the interpretation of subsurface anomalies. Here, the joint application of DC resistivity and refraction seismic tomography has shown the best results.

Acknowledgements

We would like to thank Rob McDonald, Nick Russill and Hansueli Gubler for the constructive co-laboration within the PACE work package "Geophysical Surveys", all field helpers at the various PACE field sites and the Department of Geography, University of Zurich, the glaciology section (VAW) and the Geophysical Institute, ETH Zurich for supplying the various instruments. The fieldwork was financed by the PACE project (Contract Nr ENV4-CT97-0492 and BBW Nr 97.0054-1). C. Hauck acknowledges a grant by the German Science Foundation (DFG) within the Graduiertenkolleg Natural Disasters at the University of Karlsruhe. The constructive comments of T. Pfeffer, S. Sherrard and A. Sheehan are thankfully acknowledged.

References

Annan, A. P. & Davis, J. L. (1978): High frequency electrical methods for the detection of freeze-thaw interfaces. – In: Proc. 3rd Internat. Conf. Permafrost, Edmonton. – Proc. **1**: 496–500.

Arcone, S., Lawson, D., Delaney, A., Strasser, J. C. & Strasser, J. D. (1998): Ground-penetrating radar reflection profiling of groundwater and bedrock in an area of discontinuous permafrost. – Geophysics **63** (5): 1573–1584.

Barker, R. D. (1989): Depth of investigation of collinear symmetrical four electrode arrays. – Geophysics **54** (8): 1031–1037.

Barsch, D. (1973): Refraktionsseismische Bestimmungen der Obergrenze des gefrorenen Schuttkörpers in verschiedenen Blockgletschern Graubündens. – Z. Gletscherkde. Glazialgeol. **9** (1–2): 143–167.
Benderitter, Y. & Schott, J. J. (1999): Short time variation of the resistivity in an unsaturated soil: the relationship with rainfall. – Eur. J. Environm. Engin. Geophys. **4**: 15–35.
Berthling, I., Etzelmüller, B., Isaksen, K. & Sollid, J. L. (2000): Rock glaciers on Prins Karls Forland. II: GPR soundings and the development of internal structures. – Permafrost Periglac. Process. **11** (4): 357–369.
Binley, A., Henry-Poulter, S. & Shaw, B. (1996): Examination of solute transport in an undisturbed soil column using electrical resistance tomography. – Water Resourc. Res. **32** (4): 763–769.
Daily, W., Ramirez, A., LaBrecque, D. & Nitao, J. (1992): Electrical resistivity tomography of vadose water movement. – Water Resourc. Res. **28** (5): 1429–1442.
Daniels, J. J., Keller, G. V. & Jacobson, J. J. (1976): Computer-assisted interpretation of electromagnetic soundings over a permafrost section. – Geophysics **41**: 752–765.
Doolittle, J. A., Hardiskey, M. A. & Gross, M. F. (1990): A ground-penetrating radar study of active layer thicknesses in areas of moist sedge and wet sedge tundra near Bethel, Alaska, U.S.A. – Arctic Alpine Res. **22**: 175–182.
Evin, M. & Fabre, D. (1990): The distribution of permafrost in rock glaciers of the Southern Alps (France). – Geomorphology **3**: 57–71.
Fisch, W. Sen., Fisch, W. Jun. & Haeberli, W. (1977): Electrical soundings with long profiles on rock glaciers and moraines in the Alps of Switzerland. – Z. Gletscherkde. Glazialgeol. **13** (1/2): 239–260.
French, H. K., Hardbattle, C., Binley, A., Winship, P. & Jakobsen, L. (2002): Monitoring snowmelt induced unsaturated flow and transport using electrical resistivity tomography. – J. Hydrol. **267** (3–4): 273–284.
Geonics Ltd. (1994): PROTEM47D operating manual.
Griffiths, D. H. & Barker, R. D. (1993): Two-dimensional resistivity imaging and modelling in areas of complex geology. – J. Appl. Geophys. **29**: 211–226.
Gude, M., Dietrich, S., Mäusbacher, R., Hauck, C., Molenda, R., Ruzicka, V. & Zacharda, M. (2003): Permafrost conditions in non-alpine scree slopes in central Europe. – In: Proc. 8. Internat. Conf. Permafrost, Zürich (in press).
Haeberli, W. (1973): Die Basis Temperatur der winterlichen Schneedecke als möglicher Indikator für die Verbreitung von Permafrost. – Z. Gletscherkde. Glazialgeol. **9** (1–2): 221–227.
Haeberli, W. & Vonder Mühll, D. (1996): On the characteristics and possible origins of ice in rock glacier permafrost. – Z. Geomorph. N. F., Suppl.-Bd. **104**: 43–57.
Harada, K., Wada, K. & Fukuda, M. (2000): Permafrost mapping by the transient electromagnetic method. – Permafrost Periglac. Process. **11**: 71–84.
Harris, C. & Cook, J. D. (1986): The detection of high altitude permafrost in Jotunheimen, Norway using seismic refraction techniques: an assessment. – Arctic Alpine Res. **18** (1): 19–26.
Harris, C., Haeberli, W., Vonder Mühll, D. & King, L. (2001): Permafrost monitoring in the high mountains of Europe: the PACE project in the global context. – Permafrost and Periglacial Processes **12**(1): 3–11.
Hauck, C. (2001): Geophysical methods for detecting permafrost in high mountains. – Mitt. Versuchsanst. Wasserbau, Hydrol. Glaziol. ETH Zürich **171**, 204 pp.
Hauck, C. (2002): Frozen ground monitoring using DC resistivity tomography. – Geophys. Res. Lett. **21**: 2016, doi: 10.1029/2002GL014995.
Hauck, C. & Vonder Mühll, D. (1999a): Using DC resistivity tomography to detect and characterise mountain permafrost. – In: Proc. 61. Europ. Assoc. Geoscient. Engin. (EAGE) conf., 7.–11. June 1999, Helsinki, Finland **2–15**, 4 pp.
Hauck, C. & Vonder Mühll, D. (1999b): Detecting Alpine permafrost using electro-magnetic methods. – In: Hutter, K., Wang, Y. & Beer, H. (eds.): Advances in cold regions thermal engineering and sciences. – pp. 475–482, Springer Verlag, Heidelberg.

HAUCK, C., VONDER MÜHLL, D., RUSSILL, N. & ISAKSEN, K. (2000): An integrated geophysical study to map mountain permafrost: A case study from Norway. – In: Proc. 6th EEGS meeting, 2000, Bochum, Germany, Extended Abstracts **CH01**.

HAUCK, C., GUGLIELMIN, M., ISAKSEN, K. & VONDER MÜHLL, D. (2001): Applicability of frequency-domain and time-domain electromagnetic methods for mountain permafrost studies. – Permafrost Periglac. Process. **12**(1): 39–52.

HAUCK, C., VONDER MÜHLL, D. & MAURER, H. (2003): Using DC resistivity tomography to detect and characterise mountain permafrost. – Geophysical Prospecting **51**, in press.

HOEKSTRA, P. (1978): Electromagnetic methods for mapping shallow permafrost. – Geophysics **43** (4): 782–787.

HOEKSTRA, P. & MCNEILL, D. (1973): Electromagnetic probing of permafrost. – In: Proc. 2nd Internat. Conf. Permafrost, North American Contribution, 517–526.

HOEKSTRA, P., SELLMANN, P. V. & DELANEY, A. (1975): Ground and airborne resistivity surveys of permafrost near Fairbanks, Alaska.– Geophysics **40**: 641–656.

HOELZLE, M. (1992): Permafrost occurrence from BTS measurements and climatic parameters in the Eastern Swiss Alps. – Permafrost Periglac. Process. **3**: 143–147.

HOELZLE, M., WEGMANN, M. & KRUMMENACHER, B. (1999): Miniature temperature dataloggers for mapping and monitoring of permafrost in high mountain areas: first experience from the Swiss Alps. – Permafrost Periglac. Process. **10**: 113–124.

HUBBARD, B., BINLEY, A., SLATER, L., MIDDLETON, R. & KULESSA, B. (1998): Inter-borehole electrical resistivity imaging of englacial drainage. – J. Glaciol. **44** (147): 429–434.

ISAKSEN, K., OEDEGAARD, R. S., EIKEN, T. & SOLLID, J. L. (2000): Composition, flow and development of two tongue-shaped rock glaciers in the permafrost of Svalbard. – Permafrost Periglac. Process. **11**(3): 241–257.

ISAKSEN, K., HAUCK, C., GUDEVANG, E., OEDEGAARD, R. S. & SOLLID, J. L. (2002): Mountain permafrost distribution in Dovrefjell and Jotunheimen, southern Norway, based on BTS measurements and 2D tomography data. – Norsk Geogr. Tidsskr. **56**: 122–136.

ISHIKAWA, M., WATANABE, T. & NAKAMURA, N. (2001): Genetic difference of rock glaciers and the discontinuous mountain permafrost zone in Kanchanjunga Himal, Eastern Nepal. – Permafrost Periglac. Process. **12** (3): 243–253.

KELLER, F. & GUBLER, H. (1993): Interaction between snow cover and high mountain permafrost Murtel/Corvatsch, Swiss Alps. – In: Proc. 6th Internat. Conf. Permafrost **I**: 332–337, Beijing, China.

KING, L., FISCH, W., HAEBERLI, W. & WÄCHTER, H. P. (1987): Comparison of resistivity and radio-echo soundings on rock glacier permafrost. – Z. Gletscherkde. Glazialgeol. **23** (2): 77–97.

KING, L., GORBUNOV, A. & EVIN, M. (1992): Prospecting and mapping of mountain permafrost and associated phenomena. – Permafrost Periglac. Process. **3**: 73–81.

KING, M. S., ZIMMERMAN, R. W. & CORWIN, R. F. (1988): Seismic and electrical properties of unconsolidated permafrost. – Geophys. Prospect. **36**: 349–364.

KNEISEL C., HAUCK, C. & VONDER MÜHLL, D. (2000): Permafrost below the timberline confirmed and characterized by geoelectrical resistivity measurements, Bever Valley, eastern Swiss Alps. – Permafrost Periglac. Process. **11** (4): 295–304.

KNEISEL, C. & HAUCK, C. (2003): Multi-method geophysical investigation of a sporadic permafrost occurrence. – Z. Geomorph. N. F., Suppl.-Bd. **132**: 145–159.

LANZ, E., MAURER, H. & GREEN, A. G. (1998): Refraction tomography over a buried waste disposal site. – Geophysics **63** (4): 1414–1433.

LECLAIRE, P., COHEN-TENOUDJI, F. & AGUIRRE-PUENTE, J. (1994): Extension of Biot's theory of wave propagation to frozen porous media. – J. Acoust. Soc. Amer. **96**: 3753–3768.

LEHMANN, F. & GREEN, A. G. (2000): Topographic migration of georadar data: Implications for acquisition and processing. – Geophysics **65**(3): 836–848.

LOKE, M. H. (1999): Electrical imaging surveys for environmental and engineering studies. Available in pdf-format on http://www.abem.se.

LOKE, M. H. & BARKER, R. D. (1995): Least-squares deconvolution of apparent resistivity. – Geophysics **60**: 1682–1690.

LOKE, M. H. & BARKER, R. D. (1996): Rapid least-squares inversion of apparent resistivity pseudosections using a quasi-Newton method. – Geophys. Prosp. **44**: 131–152.

MCGINNIS, L. D., NAKAO, K. & CLARK, C. C. (1973): Geophysical identification of frozen and unfrozen ground, Antarctica. – In: Proc. 2nd Internat. Conf. Permafrost: 136–146, Yakutsk, Russia.

MCNEILL, J.D. (1980): Electromagnetic terrain conductivity measurements at low induction numbers. – Technical Note, **TN-6**, Geonics Ltd.

MAIER, D., MAURER, H. & GREEN, A. G. (1995): Joint inversion of related data sets: DC resistivity and transient electromagnetic soundings. – In: Proc. 1st EEGS meeting, Turin, 461–464.

MARESCOT, L., LOKE, M. H., CHAPELLIER, D., DELALOYE, R., LAMBIEL, C. & REYNARD, E. (2003): Assessing reliability of 2D resistivity imaging in permafrost and rock glacier studies using the depth of investigation index method. – Near Surface Geophysics **1** (2), 57–67.

MUSIL, M., MAURER, H., GREEN, A. G., HORSTMEYER, H., NITSCHE, F., VONDER MÜHLL, D. & SPRINGMAN, S. (2002): Shallow seismic surveying of an Alpine rock glacier. – Geophysics **67** (6): 1701–1710.

OGILVY, R., MELDRUM, P. & CHAMBERS, J. (1999): Imaging of industrial waste deposits and buried quarry geometry by 3-D resistivity tomography. – Eur. J. Environ. Engin. Geophys. **3**: 103–113.

OLHOEFT, G.R. (1978): Electrical properties of permafrost. – Proc. 3rd Internat. Conf. Permafrost **1**: 127–131, Edmonton, Canada.

PANDIT, B. I. & KING, M. S. (1978): Influence of pore fluid salinity on seismic and electrical properties of rocks at permafrost temperatures. – Proc. 3rd Internat. Conf. Permafrost **1**: 553–566, Edmonton, Canada.

PEARSON, C., MURPHY, J., HALLECK, P., HERMES, R. & MATHEWS, M. (1983): Sonic and resistivity measurements on Berea sandstone containing tetrahydrofuran hydrates: a possible analog to natural gas hydrate deposits. – Proc. 4th Internat. Conf. Permafrost: 973–978, Fairbanks, Alaska.

RÖTHLISBERGER, H. (1972): Seismic exploration in cold region. – CRREL Monograph **II**, A2A.

ROZENBERG, G., HENDERSON, J. D. & SARTORELLI, A. N. (1985): Some aspects of transient electromagnetic soundings for permafrost delineation. – CRREL, Spec. rep. **85-5**: 74–90.

SANDBERG, S.K. (1993): Examples of resolution improvement in geoelectrical soundings applied to groundwater investigations. – Geophys. Prosp. **41**: 207–227.

SANDMEIER, K. (2002): Reflex-W, version 2.5.9. – Sandmeier Scientific Software, Karlsruhe, Germany.

SCHMÖLLER, R. & FRÜHWIRTH, R. (1996): Komplexgeophysikalische Untersuchungen auf dem Dösener Blockgletscher (Hohe Tauern, Österreich). – Beitr. Permafrostforsch. Österr., Arb. Inst. Geogr. Karl-Franzens-Univ. Graz **33**: 165–190.

SCOTT, W., SELLMANN, P. & HUNTER, J. (1990): Geophysics in the study of permafrost. – In: WARD, S. (ed.): Geotechnical and Environmental Geophysics. – pp. 355–384, Soc. Expl. Geophys.,Tulsa.

SEGUIN, M. K. (1978): Temperature-electrical resistivity relationship in continuous permafrost at Purtuniq, Ungava Peninsula. – Proc. 3rd Internat. Conf. Permafrost **1**: 137–144, Edmonton, Canada.

TELFORD, W. M., GELDART, L. P. & SHERIFF, R. E. (1990): Applied geophysics. – 2nd ed., Cambridge University Press.

TIMUR, A. (1968): Velocity of compressional waves in porous media at permafrost temperatures. – Geophysics **33** (4): 584–595.

TODD, B. J. & DALLIMORE, S. R. (1998): Electromagnetic and geological transect across permafrost terrain, Mackenzie River delta, Canada. – Geophysics **63** (6): 1914–1924.

VONDER MÜHLL, D. (1993): Geophysikalische Untersuchungen im Permafrost des Oberengadins. – Mitt. Versuchsanst. Wasserbau, Hydrol. Glaziol. ETH Zürich **122**, 222 pp.

VONDER MÜHLL, D. & SCHMID, W. (1993): Geophysical and photogrammetrical investigation of rock glacier Muragl I, Upper Engadin, Swiss Alps. – Proc. 6th Internat. Conf. Permafrost **I**: 654–659, Beijing, China.

VONDER MÜHLL, D. & KLINGELE, E. (1994): Gravimetrical investigation of ice-rich permafrost within the rock glacier Murtel-Corvatsch (Upper Engadin, Swiss Alps). – Permafrost Periglac. Process. **5**: 13–24.

VONDER MÜHLL, D., STUCKI, T. & HAEBERLI, W. (1998) Borehole temperatures in Alpine permafrost: A ten years series. – Proc. 7th Internat. Conf. Permafrost, 1085–1096, Yellowknife, Canada.

VONDER MÜHLL, D., HAUCK, C. & LEHMANN, F. (2000): Verification of geophysical models in Alpine permafrost using borehole information. – Ann. Glaciol. **31**: 300–306.

VONDER MÜHLL, D., HAUCK, C., GUBLER, H., MCDONALD, R. & RUSSILL, N. (2001a): New geophysical methods of investigating the nature and distribution of mountain permafrost with special reference to radiometry techniques. – Permafrost Periglac. Process. **12** (1): 27–38.

VONDER MÜHLL, D., DELALOYE, R., HAEBERLI, W., HOELZLE, M. & KRUMMENACHER, B. (2001b): Permafrost Monitoring Switzerland PERMOS, 1. Jahresbericht 1999/2000. – Glaziol. Komm., Schweiz. Akad. Naturwiss. SANW.

VONDER MÜHLL, D., HAUCK, C. & GUBLER, H. (2002): Mapping of mountain permafrost using geophysical methods. – Phys. Proc. Geogr. **26** (4): 640–657.

WAGNER, S. (1996): DC resistivity and seismic refraction soundings on rock glacier permafrost in northwestern Svalbard. – Norsk Geogr. Tidsskr. **50** (1): 25–36.

ZIMMERMAN, R. W. & KING, M. S. (1986): The effect of freezing on seismic velocities in unconsolidated permafrost. – Geophysics **51**: 1285–1290.

Authors' addresses: Dr. Christian Hauck, Graduiertenkolleg Naturkatastrophen, Universität Karlsruhe, Am Fasanengarten, D-76128 Karlsruhe, Germany. Dr. Daniel Vonder Mühll, Leiter Ressort Forschung, Rektorat Universität Basel, Petersgraben 35/3, CH-4051 Basel, Switzerland.

Zeitschrift für Geomorphologie   Neue Folge   Supplementbände

Piedmont plains and sand-formations — Band 10
in arid and humid tropic and subtropic regions
Ed.: H. MENSCHING. 1970. € 37,–.

Untersuchungen über Relieftypen — Band 11
arider Räume an Beispielen aus dem
Tibesti-Gebirge und seiner Umgebung
Von Prof. Dr. H. HAGEDORN. 1971. € 34,–.

Landformung durch Flüsse/ — Band 12
Geomorphological activity of streams
Hrsg.: H. LOUIS. 1971. € 45,–.

Vergriffen. — Band 13

Neue Wege der Geomorphologie/ — Band 14
A new approach to geomorphology
Zur Differenzierung der Abtragungsprozesse in verschiedenen Klimaten. Hrsg.: J. BÜDEL und C. RATHJENS. 1972. € 35,–.

Geomorphologie arider Gebiete/ — Band 15
Geomorphology of arid areas
Hrsg.: E. BIBUS und A. SEMMEL. 1972. € 37,–.

Geomorphologie des Quartärs/ — Band 16
Quaternary Geomorphology
Hrsg.: J. HÖVERMANN u. K. KAISER. 1973. € 42,–.

Beiträge zur Klimageomorphologie/ — Band 17
Contributions to climatic geomorphology
Hrsg.: H. MENSCHING und A. WIRTHMANN, Schriftleitung: H. HAGEDORN. 1973. € 48,–.

Hangforschung und Morphologie — Band 18
**der Landoberflächen/Evolution des
pentes et morphologie continentale**
Hrsg.: P. MACAR. 1973. € 40,–.

Beiträge zur regionalen Küsten- — Band 19
morphologie des Mittelmeerraumes
Von D. KELLETAT. 1974. € 38,–.

Geomorphic processes in arid environ- — Band 20
ments/Geomorphologische Prozesse arider Gebiete
Ed.: A. P. SCHICK, D. H. YAALON and A. YAIR,
Jerusalem; Schriftleitung: K. KAISER, Berlin

Volume I and II: 1974. € 49,80. — Band 21

Contributions to Coastal Geo- — Band 22
morphology/Beiträge zur Küstenmorphologie
Ed.: R. W. FAIRBRIDGE. 1975. € 45,–.

Reliefgenerationen in verschiedenen — Band 23
**Klimaten/Landform Generations in
Different Climates**
Hrsg.: J. BÜDEL und H. HAGEDORN. 1975. € 40,–.

Strukturbetonte Reliefs/ — Band 24
Landforms controlled by structure
Hrsg.: H. BLUME. 1976. € 45,–.

Quantitative slope models — Band 25
Ed.: FRANK AHNERT. 1976. € 45,–.

Karst processes — Band 26
Ed.: MAJORIE SWEETING and K.-H. PFEFFER. 1976.
€ 51,–.

Contribution to Glacial Morphology — Band 27
of the Scandinavian Ice Sheet/Zur Glazialmorphologie des skandinavischen Vereisungsgebietes
Ed.: RAJMUND GALON. 1977. € 30,–.

Hangformen und Hangprozesse/ — Band 28
Slope forms and processes
Hrsg.: ALFRED WIRTHMANN. 1977. € 51,–.

Field Instrumentation and — Band 29
Geomorphological Problems
Ed.: O. SLAYMAKER, A. RAPP and T. DUNNE. 1978.
€ 51,–.

Beiträge zur Geomorphologie der — Band 30
**Ariden Zone/Contribution to the
Geomorphology of the Arid Zone**
Hrsg.: HORST MENSCHING. 1978. € 51,–.

Inselbergs/Inselberge — Band 31
Ed.: H. BREMER and J. JENNINGS. 1978. € 51,–.

Problems in Karst Environments — Band 32
Ed.: MAJORIE SWEETING. 1979. € 30,–.

Relief und Boden — Band 33
Hrsg.: H. BREMER und H. ZAKOSEK. 1979. € 69,80.

Coast under stress — Band 34
Ed.: A. R. ORME, D. B. PRIOR, N. P. PSUTY and
H. J. WALKER. 1980. € 69,80.

Geomorphic experiment on hillslopes — Band 35
Ed.: O. SLAYMAKER, TH. DUNNE and A. RAPP. 1980.
€ 60,–.

Perspectives in Geomorphology — Band 36
Papers from the First British-German Symposium on
Geomorphology University of Würzburg 24th–
29th September 1979. Ed.: H. HAGEDORN and M.
THOMAS. 1980. € 82,–.

High Mountains/Hochgebirge — Band 37
Ed.: OLAV SLAYMAKER. 1981. € 50,–.

Reliefgenese und Paläoklima in — Band 38
**West- und Südafrika/Morphogenesis
and Paleoclimate in West and South Africa**
Hrsg.: HANNA BREMER und HORST HAGEDORN.
1981. € 29,–.

Beiträge zur angewandten Geomorpho- — Band 39
logie/Contributions in applied geomorphology
Hrsg.: HERBERT LIEDTKE. 1981. € 35,–.

Neotectonics — Band 40
Ed.: RHODES W. FAIRBRIDGE. 1981. € 65,–.

Der Dhaulagiri- und — Band 41
Annapurna-Himalaya
Ein Beitrag zur Geomorphologie extremer Hochgebirge. Von MATTHIAS KUHLE. 1982. € 99,–.

Graben – Geology and — Band 42
Geomorphogenesis
Ed.: K. BRUNNACKER and M. TAIEB. 1982. € 63,–.

Experimente und Messungen — Band 43
in der Geomorphologie
Hrsg.: DIETRICH BARSCH. 1982. € 60,–.

Applied geomorphology in the tropics — Band 44
Ed.: I. DOUGLAS and T. SPENCER. 1982. € 39,–.

Dunes: Continental and Coastal — Band 45
Ed.: J. JENNINGS and H. HAGEDORN. 1983. € 86,–.

Extreme land forming events — Band 46
Ed.: S. OKUDA, A. NETTO and O. SLAYMAKER.
1983. € 45,–.

Coastal and inland periglazial processes — Band 47
Canadian Arctic. Ed.: H. HAGEDORN. 1983. € 40,–.

Beiträge zur allgemeinen Geomor- — Band 48
**phologie der Tropen und Subtropen/
Contributions to Geomorphology
in the Tropics and Subtropics**
Hrsg.: H. ROHDENBURG und H. MENSCHING. 1983.
€ 73,–.

Present day geomorphological — **Band 49**
processes/Processus geomorphologiques actuels
Ed.: A. PISSART and J. H. J. TERWINDT. 1984. € 51,–.

Grundvorstellungen in der — **Band 50**
Geomorphologie/Concepts and Developments in Geomorphology
Vorträge des Ferdinand von RICHTHOFEN Symposiums 1983. Hrsg.: G. STÄBLEIN. 1984. € 40,–.

Applied Geomorphology — **Band 51**
Ed.: JOHN GERRARD. 1984. € 39,–.

Dambos: small channelless valleys — **Band 52**
in the tropics. – Characteristics, formation, utilisation
Ed.: M. F. THOMAS and A. GOUDIE. 1985. € 55,–.

Hypsometry of the Continents — **Band 53**
by J. GRAHAM COGLEY. 1985. € 12,–.

Morphotectonics of Passive — **Band 54**
Continental Margins
Ed.: C. D. OLLIER. 1985. € 33,–.

Fluvial Geomorphology — **Band 55**
In memoriam J. N. Jennings. Ed: HANNA BREMER. 1985. € 42,–.

Aktuelle und vorzeitliche Relief- — **Band 56**
und Bodenentwicklung
Vorträge auf dem Julius Büdel Symposium in Darmstadt 1984. Hrsg.: O. SEUFERT und H. MENSCHING. 1985. € 42,–.

Geomorphology of Changing — **Band 57**
Coastlines
Ed.: ERIC BIRD. 1985. € 50,–.

Geomorphology and — **Band 58**
Land Management
Ed.: OLAV SLAYMAKER and DAN BALTEANU. 1986. € 50,–.

UIS – Union Internationale — **Band 59**
de Spéléologie/International Atlas of Karst Phenomena. Sheets 4–7
Ed.: KARL-HEINZ PFEFFER. 1986. € 35,–.

Erosion Budgets and their — **Band 60**
Hydrologic Basis
Ed.: H. VOGT and O. SLAYMAKER. 1986. € 74,–.

Geomorphologie der Periglazialge- — **Band 61**
biete/Geomorphology of Periglacial Regions
Hrsg.: HEINZ KLUG. 1986. € 32,–.

Dating Mediterranean Shorelines — **Band 62**
Ed.: A. OZER and C. VITA-FINZI. 1986. € 57,–.

Neotectonics and Morphotectonics — **Band 63**
Ed.: C. EMBLETON. 1987. € 65,–.

Laterites — **Band 64**
Some aspects of current research
Ed.: M. J. MCFARLANE. 1987. € 50,–.

Geomorphology of European Massifs — **Band 65**
Ed.: H. BREMER and A. GODARD. 1987. € 49,–.

Geomorphologische Beiträge: — **Band 66**
Strukturformen, fluviale Dynamik, Kartierung
Hrsg.: JÜRGEN HAGEDORN. 1987. € 50,–.

Dynamic System Approach — **Band 67**
to Natural Hazards
Proceedings of a Symposium held on August 21, 1987 in Vancouver, Canada. Ed.: A. E. SCHEIDEGGER and M. J. HAIGH. 1988. € 40,–.

Applied geomorphological mapping: — **Band 68**
methodology by example
Compiled by the Working Group on Geomorphological Survey and Mapping, International Geographical Union. Ed.: C. EMBLETON. 1988. € 71,–.

The Geomorphology of Plate — **Band 69**
Boundaries and Active Continental Margins
Ed.: PAUL W. WILLIAMS. 1988. € 50,–.

Alpen und Alpenvorland — **Band 70**
Beiträge zur Geomorphologie
Hrsg.: D. BARSCH und M. GAMPER. 1988. € 69,–.

Periglacial processes and landforms — **Band 71**
Ed.: EDUARD A. KOSTER and HUGH M. FRENCH. 1988. € 60,–.

Weathered mantles (saprolites) over — **Band 72**
basement rocks of High Latitudes/ Les manteaux d'altération sur les socles des Hautes Latitudes
Ed.: A. GODARD. 1989. € 63,–.

Coasts: Erosion and sedimentation — **Band 73**
Ed.: H. BREMER and K. M. CLAYTON. 1989. € 64,–.

Beiträge zur Geomorphologie Afrikas — **Band 74**
Hrsg.: K. HÜSER und H. STINGL. 1989. € 55,–.

Karst — **Band 75**
Ed.: K.-H. PFEFFER. 1989. € 45,–.

Quaternary of the Karakoram — **Band 76**
and Himalaya
Ed.: E. DERBYSHIRE and L. A. OWEN. 1989. € 79,–.

Sheets 8–12. International Atlas — **Band 77**
of Karst Phenomena/Union Internationale de Spéléologie
Ed.: KARL-HEINZ PFEFFER. 1990. € 40,–.

Climate Related Landscapes in — **Band 78**
World Mountains: Criteria and Map
By WILL F. THOMPSON. 1990. € 40,–.

Proceedings of the 2nd International Conference on Geomorphology: Geomorphology and Geoecology, Frankfurt/Main 1989 = Band 79–88

Vol. I: General, Invited and — **Band 79**
Special Lectures
Ed.: D. BARSCH. 1990. € 60,–.

Vol. II: Geomorphological mapping, — **Band 80**
remote sensing and terrain models
Ed.: C. EMBLETON and H. LIEDTKE. 1990. € 25,–.

Vol. III: Coastal Dynamics and — **Band 81**
Environments
Ed.: R. P. PASKOFF and D. KELLETAT. 1991. € 60,–.

Vol. IV: Morphotectonics — **Band 82**
and Structural Geomorphology
Ed.: C. D. OLLIER. 1991. € 50,–.

Vol. V: Geomorphological — **Band 83**
Approaches in Applied Geography
Ed.: J. DE PLOEY, G. HAASE and H. LESER. 1991. € 81,–.

Vol. VI: Climatic Geomorphology — **Band 84**
Ed.: H. HAGEDORN and A. RAPP. 1992. € 63,–.

Vol. VII: Karst — **Band 85**
Ed.: J. NICOD, J.-H. PFEFFER and M. SWEETING. 1992. € 42,–.

Vol. VIII: Glacial and Polar — **Band 86**
Geomorphology
Ed.: G. STÄBLEIN, H. M. FRENCH and M. G. MARCUS. 1992. € 55,–.

Vol. IX: Applied Geomorphology — **Band 87**
Ed.: M. PÉCSI and G. RICHTER. 1993. € 50,–.

Vol. X: **Fluvial Geomorphology** — Band 88
Ed.: I. Douglas and J. Hagedorn. 1993. € 35,–.

Aktuelle Geomorphodynamik und angewandte Geomorphologie – Present-day Geomorphodynamics and Applied Geomorphology — Band 89
Hrsg.: R. Mäckel. 1991. € 55,–.

Late Vistulian (= Weichselian) and Holocene Aeolian Phenomena in Central and Northern Europe — Band 90
Ed.: S. Kozarski. 1991. € 76,–.

Geomorphology of the Tropics with special reference to South Asia and Africa — Band 91
Ed. J. Grunert. 1992. € 74,–.

Some Contributions to the Study of Landforms and Geomorphic Processes — Band 92
edited on behalf of Deutscher Arbeitskreis für Geomorphologie by D. Barsch and R. Mäusbacher. 1993. € 76,–.

Klimagenetische Geomorphologie Climato-Genetic Geomorphology — Band 93
Hrsg.: R. W. Fairbridge and K.-H. Pfeffer. 1993. € 65,–.

Neotectonics and active faulting — Band 94
edited by I. Stewart, C. Vita-Finzi and L. Owen. 1993. € 99,80.

Last Ice Sheet Dynamics and Deglaciation in the North European Plain — Band 95
edited by M. Böse and S. Kozarski. 1994. € 45,–.

Periglaziale Deckschichten und Böden im Bayerischen Wald und seinen Randgebieten als geogene Grundlagen landschaftsökologischer Forschung im Bereich naturnaher Waldstandorte — Band 96
von Jörg Völkel. 1995. € 76,–.

Geowissenschaftliche Spitzbergen-Expedition 1990–1992 (SPE 90–92) – Liefde-, Wood- und Bockfjord/NW-Spitzbergen — Band 97
Hrsg.: Wolf Dieter Blümel. 1994. € 91,–.

Löss: Herkunft, Gliederung, Landschaften — Band 98
Hrsg.: M. Pécsi and G. Richter. 1996. € 76,–.

Form and Process in Geomorphology — Band 99
edited by K. Heine, H. Strunk and J. Völkel. 1995. € 43,–.

Late Quaternary and present-day fluvial processes in Central Europe. — Band 100
edited by J. Hagedorn. 1995. € 63,–.

Advances in Geomorphometry – Proceedings of the Walter F. Wood Memorial Symposium — Band 101
edited by R. J. Pike and R. Dikau. 1995. € 71,–.

Field Methods and Models to Quantify Rapid Coastal Changes — Band 102
edited by D. H. Kelletat and N. P. Psuty. 1996. € 71,–.

Tropical/Subtropical Geomorphology – Research studies from coastal areas to high mountains — Band 103
edited by K.-H. Pfeffer. 1996. € 96,–.

High Mountain Geomorphology — Band 104
edited by R. Mäusbacher and A. Schulte. 1996. € 60,–.

Recent Research in Austalasian Geomorphology — Band 105
edited by I. F. Owens. 1996. € 60,–.

Weathering – Erosion – Sedimentation — Band 106
edited by K.-H. Pfeffer. 1996. € 76,–.

Soil erosion and land degradation in regions of Mediterranean climate — Band 107
edited by M. E. Meadows and M. Sala. 1996. € 32,–.

Tropical and Subtropical Karst – essays dedicated to the memory of Dr. Marjorie Sweeting — Band 108
edited by P. W. Williams. 1997. € 33,–.

Sheets 13–17. International Atlas of Karst Phenomena/Union Internationale de Spéléologie — Band 109
Ed.: Karl-Heinz Pfeffer. 1998. € 49,–.

Geomorphology and changing environments in Central Europe — Band 110
edited by H. Bremer and D. Lóczy. 1997. € 65,–.

Eolian Dynamics: Landforms and processes — Band 111
edited by Wolf Dieter Blümel. 1997. € 58,–.

Abbildung von Prozessen und Strukturen in Geosystemen — Band 112
hrsg. von K. D. Aurada, K. Billwitz und R. Lampe. 1998. € 70,–.

Vergletscherungen in europäischen Mittelgebirgen — Band 113
hrsg. von A. Kostrzewski und H. Hagedorn. 1999. € 35,–.

Proceedings of the Fourth International Conference on Geomorphology, Bologna 1997. Vol. I: Volcanic Geomorphology — Band 114
edited by P. Fredi. 1999. € 38,–.

Proceedings of the Fourth International Conference on Geomorphology, Bologna 1997. Vol. II: Magnitude and Frequency in Geomorphology — Band 115
edited by R. Mäusbacher. 1999. € 49,–.

Aeolian Geomorphology – Papers from the 4th International Conference on Aeolian Research, Oxford, UK, 1998 — Band 116
edited by I. Livingstone. 1999. € 61,–.

Reliefentwicklung, Pedogenese und geoökologische Probleme agrarischer Nutzung eines tropischen Berglandes – das Beispiel Nordthailand — Band 117
edited by J. Kubiniok. 1999. € 48,–.

Proceedings of the Fourth International Conference on Geomorphology, Bologna 1997. Vol. III: Tectonic Geomorphology — Band 118
edited by W. Frisch. 1999. € 71,–.

Proceedings of the Fourth International Conference on Geomorphology, Bologna 1997. Vol. IV: Weathering and Soils — Band 119
edited by C. F. Pain. 1999. € 39,80.

Recent advances in field and laboratory studies of rock weathering — Band 120
edited by Heather A. Viles. 2000. € 65,–.

Geomorphologische Prozeßforschung. Stofftransport, Methodik und Regionale Aspekte — Band 121
Hrsg. von R. Mäusbacher, J. Baade und M. Gude. 2000. € 60,–.

Holocene Geomorphology Band 122
edited by K.-H. Pfeffer. 2000. € 86,–.

Angewandte und vernetzte Band 123
geomorphologische Prozeßforschung
Hrsg. im Namen des Deutschen Arbeitskreises für Geomorphologie e.V. von M. Becht und K.-H. Schmidt. 2000. € 39,80.

Angewandte Geomorphologie Band 124
in verschiedenen Geoökosystemen
Hrsg. von K. Heine u. K.-H. Pfeffer. 2001. € 56,–.

Mass movements in South, West Band 125
and Central Germany
edited by R. Dikau and K.-H. Schmidt. 2001. € 66,–.

Research in Deserts and Mountains Band 126
of Africa and Central Asia
edited by F. Lehmkuhl, D. Busche and B. Wünnemann. 2002. € 74,–.

Late Quaternary Band 127
Geomorphodynamics
edited by K.-H. Schmidt and Th. Vetter on behalf of the Deutscher Arbeitskreis für Geomorphologie. 2002. € 74,–.

Environmental change Band 128
and geomorphology
edited by R. Baumhauer and B. Schütt. 2002. € 79,–.

South and Central American Rivers Band 129
edited by J. Mossa, E. Latrubesse and A. Gupta. 2002. € 59,–.

Glaciation and Periglacial Band 130
in Asian High Mountains
Proceedings of the 5th International Conference on Geomorphology, Tokyo, Japan, August 23–28, 2001 edited by M. Böse, K. Hirakawa, N. Matsuoka and T. Sawagaki. 2003. € 98,–.

Karst in a changing world Band 131
edited by K. Urushibara-Yoshino and P. Williams. 2003. € 58,–.

Die Reihe wird fortgesetzt

Gebrüder Borntraeger · Berlin · Stuttgart

Hirt^s Stichwortbücher

Wilhelmy, Herbert:

Geomorphologie in Stichworten. II. Exogene Morphodynamik

Abtragung – Verwitterung – Tal- und Flächenbildung

6. überarb. Auflage von **Berthold Bauer** und **Hans Fischer**.
2002. X, 203 Seiten, 58 Abb., 19 x 13 cm, ISBN 3-443-03113-7, brosch., € 19.80

Dem Band I der "Geomorphologie in Stichworten", der die endogenen Kräfte, Vorgänge und Formen behandelt, folgt nunmehr in 6. Auflage Teil II, der ebenso wie Teil III die exogenen Kräfte, Vorgänge und Formen zum Thema hat.
Der Umfang des Stoffs machte die Behandlung der exogenen Morphodynamik in 2 Teilen erforderlich, von denen der vorliegende Band sich mit Fragen der Verwitterung, den Böden als Indikatoren morphologischer Prozesse, den Formen der Abtragung und ihrer Ergebnisse beschäftigt. Nach allgemeinen Erörterungen über die spezifischen Vorgänge der Erosion und Denudation werden in ausführlicher Darstellung deren Resultate untersucht: Täler und Flächen, in die sich letztlich das gesamte Relief der Erde gliedert.
Die nun vorliegende 6. Auflage wurde von Hans Fischer und Berthold Bauer überarbeitet. Einige Kapitel sind völlig neu gestaltet bzw. stark überarbeitet worden, wie z.B. Übersicht über die exogenen Kräfte, Verwitterung, Prozesse der Hangformung, Flußwerk; am Ende des Bandes wurde ein Abschnitt "Fachbegriffe" neu eingeführt. Abgegangen wurde von dem Prinzip, nach jedem Abschnitt die einschlägige Literatur darzubieten, dafür wurde ein umfassendes Literaturverzeichnis neu zusammengestellt, wobei vor allem das neuere Schrifttum berücksichtigt wurde.

Gebrüder Borntraeger · Berlin · Stuttgart

Auslieferung: E. Schweizerbart'sche Verlagsbuchhandlung (Nägele u. Obermiller), Johannesstr. 3A, D-70176 Stuttgart, Tel. +49-(0)711-625001, Fax +49-(0)711-625005
mail@schweizerbart.de   http://www.schweizerbart.de

Short description of the CD with the title:

Sandmeier Software
Defraction seismics Demo
REFLEXW Version 3.0
Copyright 1993–2003 by
K. J. Sandmeier, Karlsruhe, Germany
(=Addendum to Zeitschrift für Geomorphologie, Supplement Vol. 132:
Geophysical Applications in Geomorphology)

The enclosed CD contains a demo version of the program REFLEXW for the 2- and 3-dimensional processing and interpretation of reflection, refraction and transmission data with a wide range of applications: GPR (Ground Penetrating Radar), reflection seismics, refraction seismics, borehole transmission, ultrasound.

Apart from the complete range of the standard filter- and CMP-processing steps many elements especially designed for various applications are incorporated (e.g. 3D-data-interpretation incl. calculation of timeslices, picking, wavefront-inversion of first arrival traveltimes, raytracing, simulation of the wave propagation using a Finite Difference (FD) method and tomographic interpretation of the traveltimes based on SIRT).

The CD includes the demo-program with demodata as well as a short introduction into the seismic refraction interpretation using REFLEXW and a short technical note about the use of the different seismic refraction interpretation tools.